ATLAS
OF
HUMAN
EVOLUTION

ATLAS OF HUMAN EVOLUTION Second Edition

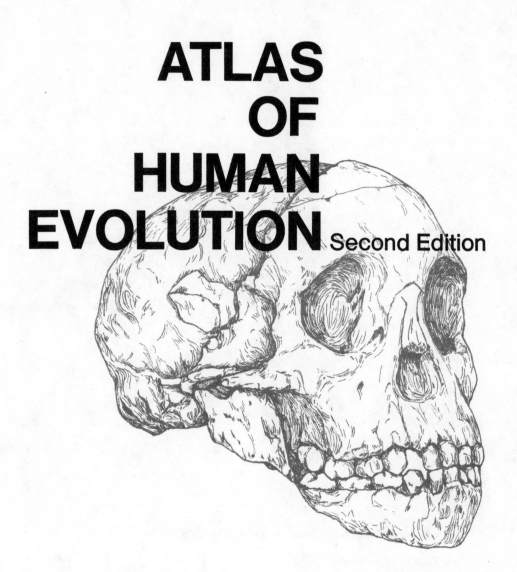

C. LORING BRACE *University of Michigan*

HARRY NELSON *Foothill College*

NOEL KORN *Los Angeles Valley College*

MARY L. BRACE *Illustrator*

HOLT, RINEHART AND WINSTON
New York   Chicago   San Francisco   Dallas
Montreal   Toronto   London   Sydney

Cover and title page skull is "Taung."

Library of Congress Cataloging in Publication Data

Main entry under title:
Atlas of human evolution.
    Brace, C. Loring

    First ed. (1971) by C. L. Brace,
H. Nelson, N. Korn, published title:
Atlas of fossil man.
    Bibliography: p. 175
    Includes index
    1. Fossil man.   I.   Brace, C. Loring
Atlas of human evolution.

GN282.B7   1979   573'.3   78–27723
ISBN 0–03–045021–7

occasionally in addition to) the standard anatomical full face or side views. This, we feel, gives the viewer a more three-dimensional impression and allows him to assess the relationship between face and braincase development which has undergone important changes in the course of human evolution. The orientation of skulls and cranial fragments is, except when noted, on the so-called Frankfurt Plane, that is, along an imaginary horizontal plane passing through the lower margin of the eye orbit and the upper margin of the ear opening.

Finally we should comment on the fact that our atlas is limited almost entirely to a portrayal of skulls, jaws, and teeth. There are two reasons for this. First, most of the evidence for human evolution exists only in the form of skulls, jaws, and teeth. Second, almost all of the major changes that have taken place during the course of human evolution are changes in craniofacial form. Some changes in pelvic form have occurred as bipedal locomotion was perfected, and we have illustrated them insofar as the very limited evidence allows. Most of our illustrations, however, depict skulls, jaws, and teeth where the evidence is most abundant and where changes have been most dramatic.

In the notes accompanying each drawing, DESIGNATION refers to the names most commonly encountered in the literature and the scientific name; DATING is the geological age accepted by the present authors, not always in agreement with dates supported by others; DISCOVERED BY refers to the original discovery of the type specimen and occasionally to the discoverer of one or more other important specimens; LOCALE lists the original site in addition to sites of the most important fossils assigned to the DESIGNATION; and SOURCE lists the original descriptions as well as selected studies of the original fossil materials. COMMENT stands for itself, or better, for what we feel is important to note for each of the specimens pictured.

Since the material was assembled for the first edition of this atlas nearly a decade ago, the pace of discovery has quickened. Exciting new finds have been made from western Europe to Java and at various points in between. Most momentous of all, however, has been the profusion of very early hominid fossils from East Africa— Laetoli in Tanzania, Koobi Fora in Kenya, and Hadar in Ethiopia. One sometimes hears it said that these have caused a revolution in our ideas concerning human origins. If this is a bit overly melodramatic, it is nonetheless true that the new African discoveries have helped us to understand the broad outlines of early hominid evolution while, tantalizingly, confronting us with problems that can only be solved by still more finds yet to be made. In the pictures and comments that follow, we have tried to keep the reader as up-to-date as it is possible to be on the basis of what has been published up to the beginning of 1979.

C. L. B.
H. N.
N. K.
M. L. B.

PREFACE

This atlas is designed for those students, beginning or advanced, professional or not, who wish to have a visual picture of the more important pieces of the evidence for human evolution. Many claims, often contradictory, have been made about various specimens, and it is our feeling that students should have the right to see the evidence for themselves. Not that one or two drawings are sufficient for even the professional scholar to reach a decision, but verbal assertions in the absence of any visual presentation are even less satisfactory. In this matter we reaffirm the old proverb that a picture is worth a thousand words.

On the other hand, unlabeled pictures are of even less value than unillustrated descriptions. Consequently, even though our aim is to provide a pictorial presentation, we include enough verbal information so that readers can place the specimen in context, so that they can check the most important publications should they so desire, and so that they have some impression of the significant features and their interpretations.

Many drawings in well-known texts and monographs are quite often inaccurate in subtle and misleading ways. This is due to the unconscious but inevitable distortion on the part of the artist or illustrator. In our atlas, however, all of the drawings have been prepared with the aid of a camera lucida and the result is that the contours and features of the objects represented are as free as possible from unconscious misrepresentation. Most of the drawings have been made from photographs, although some have been made from casts. Where the specimen portrayed has been reconstructed, we have so indicated by stippling.

This collection should not be regarded as an attempt to illustrate all of the evidence for human evolution, much of which is highly fragmentary in nature. Rather we have chosen the best known and most complete specimens. Where possible, we have included more than one specimen of the major stages of human evolution with an eye toward illustrating the range of variation among the populations in question. Our feeling is that just as no one individual would be adequate to represent the totality of modern men and women so it was in the past. Where possible, also, we have shown individuals from the same time level but from different portions of the world.

The reader will further note that in many instances we show more than one view of the same specimen. Only rarely does a single view present all the most important features, and we feel that some of the most significant pieces yield valuable perspectives when seen in more than a single pose. We also want to draw attention to the fact that we have often used a three-quarters view rather than (and

ACKNOWLEDGMENTS

The fossils omitted as much as those included and the comments not made as much as those expressed reflect the influence of a great many people both before and during the time that this atlas was being compiled. Of course, however, the decisions behind these determinations were ours alone. Of particular importance in shaping our views have been discussions with Prof. H. B. S. Cooke, Dalhousie University, Nova Scotia; Dr. J.-L. Heim, Musée de l'Homme, Paris; Prof. F. C. Howell, University of California at Berkeley; Mr. A. R. Hughes, University of the Witwatersrand, Johannesburg, South Africa; Prof. A. J. Jelinek, University of Arizona; the late Prof. T. D. McCown, University of California, Berkeley; the late Prof. G. K. Neumann, University of Indiana; Prof. J. T. Robinson, University of Wisconsin; Prof. A. C. Walker, Johns Hopkins University; Prof. T. D. White, University of California, Berkeley; and Prof. M. H. Wolpoff, University of Michigan. More specific help in the form of photographs of, or access to, original specimens was given by Dr. C. K. Brain, Transvaal Museum, Pretoria, South Africa; Dr. D. R. Brothwell, British Museum (Natural History); the late Dr. A. C. Hoffman, National Museum, Bloemfontein, South Africa; Prof. W. W. Howells, Harvard University; Dr. D. C. Johanson, Cleveland Museum of Natural History; Dr. G. H. R. von Koenigswald, Senckenburg Museum, Frankfurt, Germany; the late Dr. L. S. B. Leakey, Dr. M. D. Leakey and Mr. R. E. F. Leakey, National Museum, Nairobi, Kenya; the late Dr. J. Poljak and Dr. I. Crnolatac, National Geological and Paleontological Museum, Zagreb, Yugoslavia; Dr. Hamo Sassoon, University of Khartoum, Khartoum, Sudan; E. M. Shaw and Q. B. Hendey, South African Museum, Cape Town, South Africa; Prof. E. L. Simons, Duke University; Prof. H. Suzuki, University of Tokyo; and Prof. P. V. Tobias, University of the Witwatersrand, Johannesburg, South Africa.

Expert assistance on specific topics was provided by colleagues of the senior author at the Museum of Anthropology, University of Michigan; particularly Prof. W. R. Farrand, Prof. P. D. Gingerich and Prof. J. D. Speth. Special gratitude is tendered to Prof. *emeritus* J. B. Griffin, former Director of the Museum of Anthropology, University of Michigan, since his efforts were instrumental in providing the funds which enabled the senior author to study many of the specimens portrayed.

The problem of visiting sites and studying the crucial specimens in the Transvaal region of South Africa with a minimum of allotted time was greatly facilitated by the hospitality, contacts, and chauffeur service so graciously and generously given by Eunice R. Broido of Johannesburg. Most valuable of all have been the services of

Dr. Jeffrey Cossman of the University of Michigan Medical School. Without his cooperative, understanding, and able assistance in the darkroom, the laboratory, with occasionally recalcitrant museum officials and with a host of other complications, much of this atlas could never have been done. We also extend our thanks to Margot Massey for suggestions concerning the layout and the appearance of the cover.

Finally, we should acknowledge the extraordinary efforts needed to produce the drawings which form the core of our atlas. *Dryopithecus,* the occlusal view of the "Taung" mandible, the side view of "Saldanha," and the occlusal view of the "Gibraltar" skull were done by Vicki Jennings. The rest were drawn by Mary L. Brace. The task of the artists was frequently complicated by the authors' insistence that a specimen appear in a particular pose or that a certain detail was of greater importance than another and should be drawn as such. If the reader finds fault with the art work, then the blame should be placed squarely upon the authors who stubbornly insisted that the specimens appear just as they do— often to the dissatisfaction of the artist. Now that the work is done, we apologize for our intransigence. Here, too, we wish to express our pleasure at the graphic results. We hope that the labors necessary for their production shall have been worthwhile and that the atlas will prove truly useful in displaying the crucial steps which have occurred in the course of human evolution so far. Last but not least, we should like to record our sincere appreciation for the encouragement, faith, and patience of our editor, David P. Boynton of Holt, Rinehart and Winston.

CONTENTS

INTRODUCTION

Pliocene-Pleistocene Climate and Chronology

The evolution of the genus *Homo* took place during the geological period called the Pleistocene, once conceived of as being the "Ice Age." Later, geologists identified and named four episodes of intensified cold during the Pleistocene. Then the Yugoslav astronomer Milutin Milanković predicted that the earth should have experienced repeated instances of chilling and warming that recurred at intervals of approximately 100,000 years. Now more than fifty years later, it is clear that Milanković was absolutely right.

The orbit taken by the earth as it revolves around the sun varies from circular to slightly elliptical and back again, and it takes about 100,000 years for the cycle to be accomplished. During those times when the orbit is most nearly circular, there is no part of the year when the earth is less than its average distance from the sun. This is when the yearly temperature average drops, and, at high latitudes, the summers are not warm enough to melt all of the previous winter's snow. Each year sees the addition of a little more unmelted snow and the result is the formation of snow fields that become compacted to form continental glaciers of over a mile in thickness. Repeated changes in the tilt of the axis of daily rotation also contribute to the effect created by changes in the shape of the orbit so that within each 100,000 year cycle there are spells of intensification of cold that alternate with times of partial amelioration.

Dating the events of the past can occasionally be done directly, but more often it is accomplished by indirect methods. The decay of radioactive elements proceeds at a fixed rate after they have been laid down in the crystals where they are found. For example, some of the carbon incorporated when bone crystals are formed is the radioactive isotope carbon 14. As time goes on, this changes—decays—at a fixed rate into the stable isotope carbon 12. The proportion of ^{14}C to ^{12}C can tell the analyst how many years it has been since carbon was being incorporated into the bone in question, that is, how long since the bone was part of a living creature. Refinements of analysis now allow us to use ^{14}C back to about 70,000 years.

Beyond that, however, we are forced to rely on other methods. Volcanic rocks are laid down with set proportions of radioactive and stable isotopes. As the radioactive isotopes decay, the proportions change and allow us to say how long it has been since the rock was formed. Some of the radioactive isotopes in volcanic rocks have very slow rates of decay and can be used for age determinations on the order of millions or even billions of years. Animal bones that are in a geological layer just beneath an ancient lava flow must be at least as old as the date determined for the lava. Among the more useful indicators for the age of ancient rocks are the proportions between rubidium 87 and strontium 87, uranium 238 and lead 206, proactinium 231 and thorium 230, and potassium 40 and argon 40 (K/Ar).

Another less direct but still very useful indicator is the polarity status of ancient rocks. The earth is a huge although weak magnet which is why the needle on a compass orients on a north-south axis. Periodically in the past, the north and south magnetic poles of the earth have been reversed. We do not know what triggers the change, but, at intervals of roughly a million years, the mode of polarity is the reverse of what it had been for the preceding million-year epoch. Within each epoch there are a series of minor "events" where the polarity reverses itself again for brief stretches of time. As volcanics and marine sedimentary rocks are laid down, their particles orient to the prevailing polarity. Even in the absence of a radiometric date, the polarity profile of a sequence of prehistoric sediments can be matched against a sequence elsewhere that has such a date. In this way, dates can be established for many fossil-bearing prehistoric strata.

The chart opposite shows the events of the geological time scale since the beginning of the Pliocene. The first hominids appear in the Middle Pliocene. Stone tools appear at the border between the Pliocene and the Pleistocene at a time when the hundred-thousand-year glacial-interglacial oscillations begin. About a dozen of these occurred before the Middle Pleistocene began with the onset of the Brunhes Normal Epoch (our present polarity) was established 700,000 years ago. Eight subsequent glacial-interglacial oscillations have occurred, and there is every indication that we are now approaching the end of an interglacial, and that the next glaciation is due to occur in another thousand years or so.

SOURCES: Nigel Calder, 1975, *The Weather Machine.* New York: The Viking Press, 143 pp.

Stuart Fleming, 1977, *Dating in Archaeology.* New York: St. Martin, 272 pp.

Murray J. Mitchell, Jr., 1977, The changing climate. In Geophysics Study Committee (ed.), *Energy and Climate*, Washington, D.C.: National Academy of Sciences, pp. 51–58.

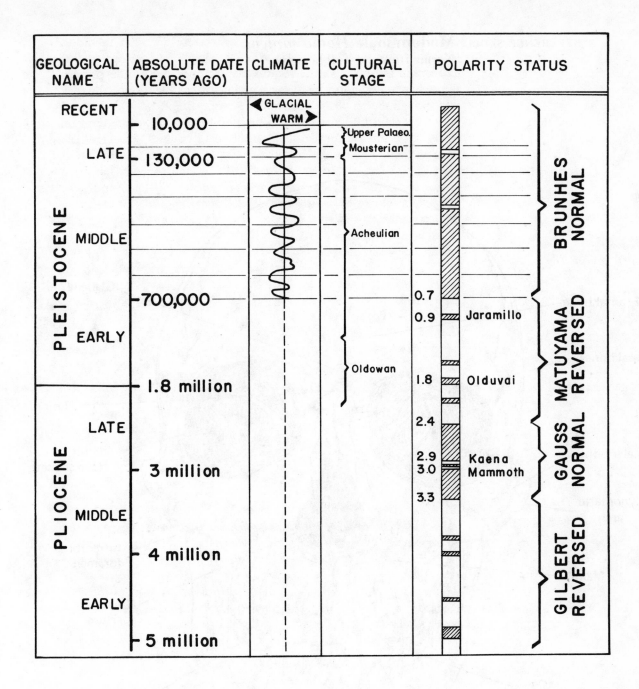

GEOLOGICAL NAME	ABSOLUTE DATE (YEARS AGO)	CLIMATE	CULTURAL STAGE	POLARITY STATUS
RECENT	10,000	GLACIAL ◄ WARM ►	Upper Palaeo.	BRUNHES NORMAL
PLEISTOCENE LATE	130,000		Mousterian	
PLEISTOCENE MIDDLE			Acheulian	
PLEISTOCENE EARLY	700,000		Oldowan	0.7 MATUYAMA REVERSED; 0.9 Jaramillo
	1.8 million			1.8 Olduvai
PLIOCENE LATE	3 million			2.4; 2.9 Kaena; 3.0 Mammoth GAUSS NORMAL
PLIOCENE MIDDLE	4 million			3.3 GILBERT REVERSED
PLIOCENE EARLY	5 million			

3

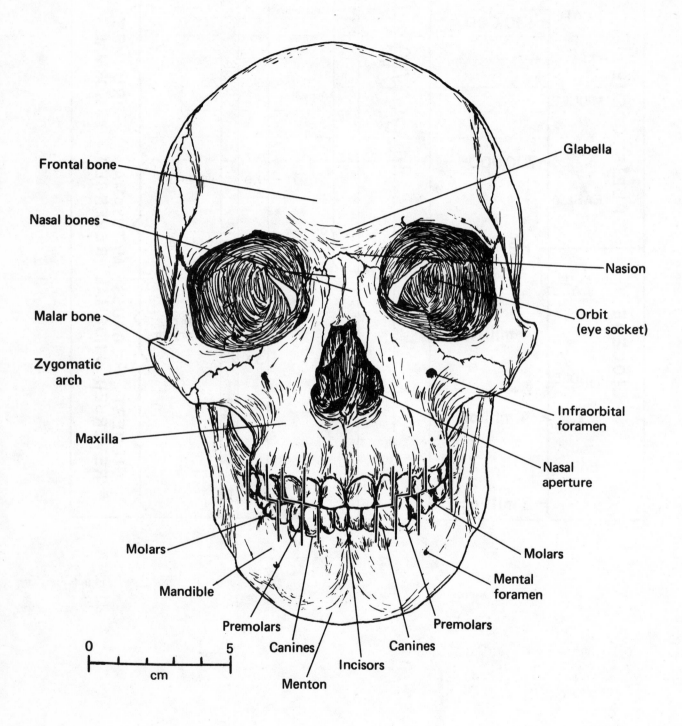

4

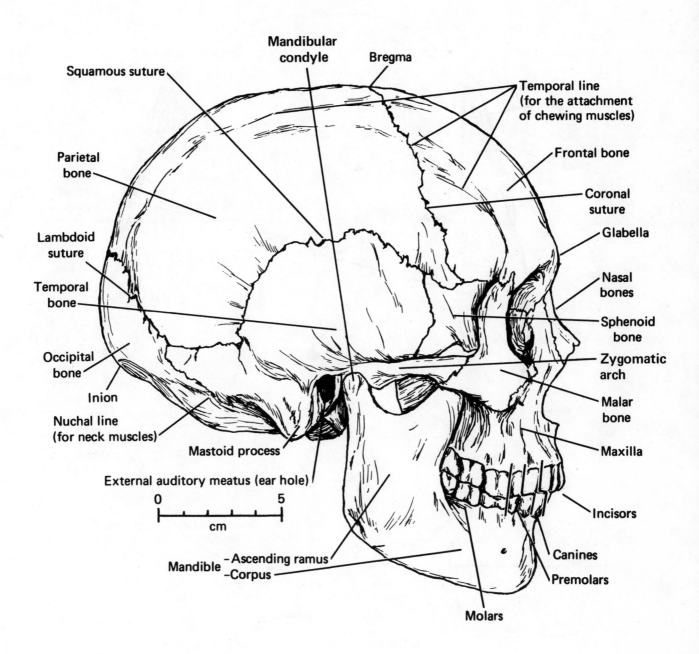

Squamous suture

Mandibular condyle

Bregma

Temporal line (for the attachment of chewing muscles)

Parietal bone

Frontal bone

Coronal suture

Lambdoid suture

Glabella

Temporal bone

Nasal bones

Sphenoid bone

Occipital bone

Zygomatic arch

Inion

Malar bone

Nuchal line (for neck muscles)

Mastoid process

Maxilla

External auditory meatus (ear hole)

0   5

cm

Incisors

Mandible – Ascending ramus
          – Corpus

Canines

Premolars

Molars

REFERENCE SKULL: **Modern male** *Homo sapiens*
Norma verticalis (top view)

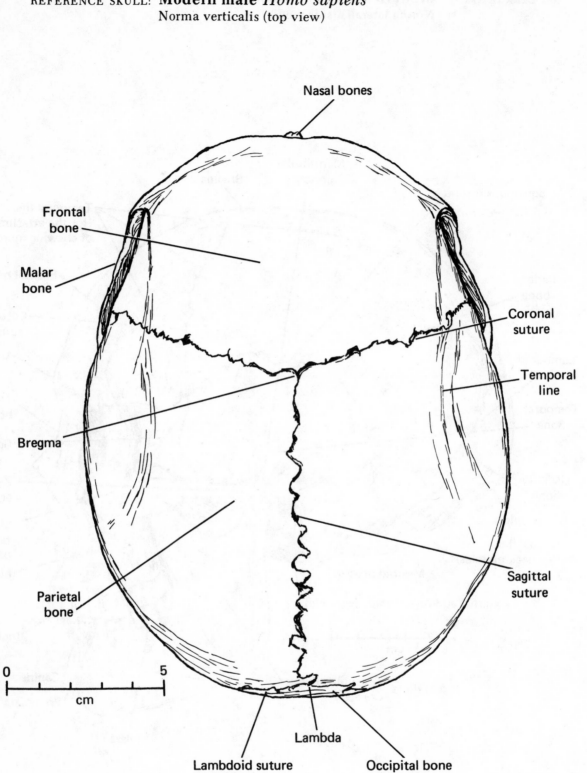

Nasal bones

Frontal
bone

Malar
bone

Coronal
suture

Temporal
line

Bregma

Sagittal
suture

Parietal
bone

0   5
cm

Lambda

Lambdoid suture

Occipital bone

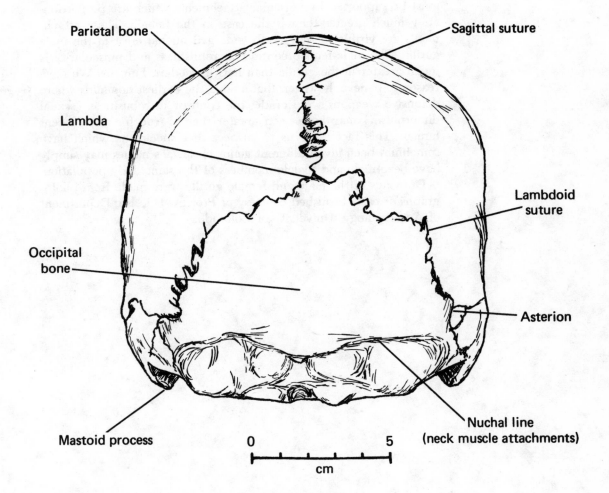

Parietal bone

Sagittal suture

Lambda

Lambdoid suture

Occipital bone

Asterion

Nuchal line (neck muscle attachments)

Mastoid process

0   5

cm

DESIGNATION: *Gorilla gorilla*
Female as compared with male.

COMMENT: The female gorilla lacks the extreme development of the brow ridge which is found in the male. The female also lacks the prominent development of the sagittal crest which is characteristic of the male. In the male this crest provides a much larger area for the attachment of the jaw muscles which completely encase the skull and rise up to the top of the crest. With the enlarged canine teeth and extensive jaw muscles, the male gorilla is capable of delivering a truly formidable bite. Wielding the dentition as a major weapon at the forward end of the face and skull means that the whole head must be supported by muscular attachments which are proportionately much greater than is the case in the female. These attachments are visible in the greatly enlarged nuchal crest in the male gorilla. The male-female contrast in robustness and muscularity is much greater in the gorilla than it is in modern humans. We have reason to believe, however, that among the earliest hominids where defensive weapons were crude, the contrast in robustness (sexual dimorphism) may have been greater than is true for more recent humans.This fact leads us to suspect that specimens which have sometimes been given different genus or species names may simply have been male and female members of the same early population.

Drawings of the male and female gorilla were made from photographs of casts furnished courtesy of Prof. A. J. Kelso, Department of Anthropology, University of Colorado.

Adult Male Gorilla

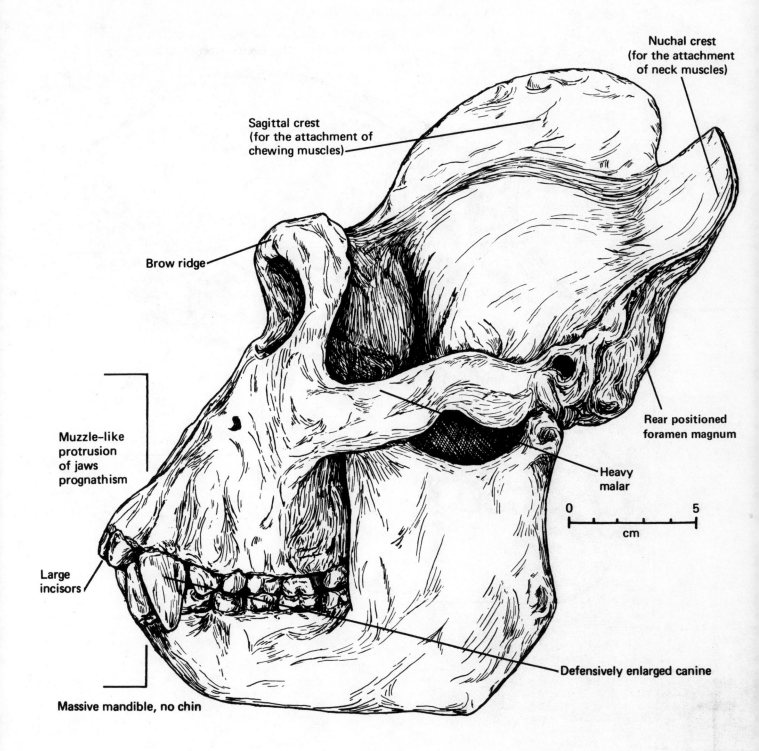

Nuchal crest
(for the attachment
of neck muscles)

Sagittal crest
(for the attachment of
chewing muscles)

Brow ridge

Muzzle–like
protrusion
of jaws
prognathism

Rear positioned
foramen magnum

Heavy
malar

Large
incisors

Defensively enlarged canine

Massive mandible, no chin

0   5
cm

Adult Female Gorilla (no sagittal crest)

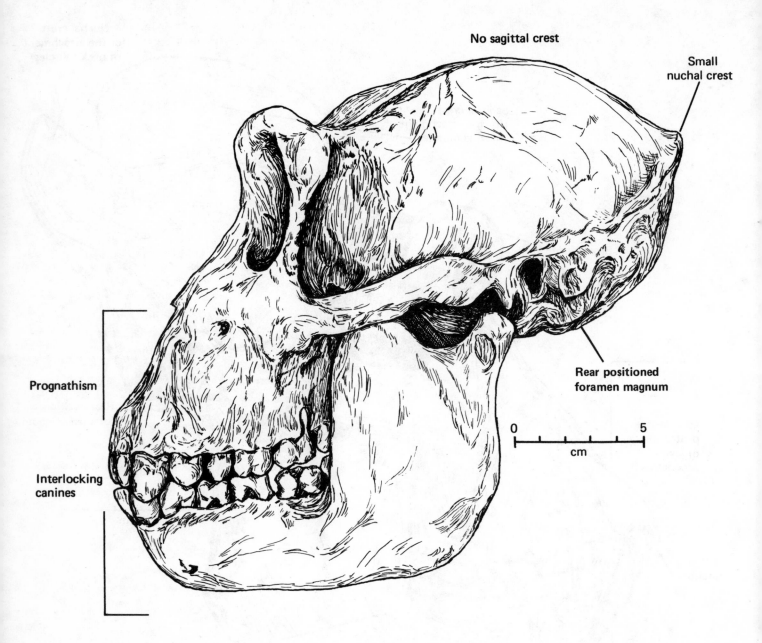

No sagittal crest

Small
nuchal crest

Rear positioned
foramen magnum

Prognathism

Interlocking
canines

0   5

cm

Taxonomy

Taxonomy is the science of naming. Its application results in the classification of living things. Descending the taxonomic hierarchy, each category represents a more exclusive (and often more recent) evolutionary relationship. A simplified classification for man shows the name of each taxonomic category followed by the name of the division to which humans belong:

KINGDOM	Animalia
PHYLUM	Chordata
CLASS	Mammalia
ORDER	Primates
FAMILY	Hominidae
GENUS	*Homo*
SPECIES	*sapiens*

Many of these categories can be further divided. For instance, a "superfamily" is more inclusive than a family but less inclusive than an order. As an example, the order Primates includes prosimians (lemurs and lorises), monkeys, apes, and humans. The superfamily Hominoidea includes families for apes and people but not for monkeys and prosimians. The fossil and living great apes belong to family Pongidae which we can refer to as "pongids," and fossil and living humans belong to family Hominidae—informally, "hominids."

Following is a possible Primate family tree. Family trees are never more than educated guesses suggesting possible evolutionary relationships. This is our guess based on the somewhat skimpy fossil record plus the chemical and morphological similarities and differences in the living Primates. Note that the Primates as an order became distinct from the other mammals (represented by the primitive order Insectivora) by the beginning of the Cenozoic Era (often called the Age of Mammals). Pongids became recognizable by the late Oligocene and hominids perhaps in the late Miocene but certainly in the Pliocene.

PRIMATE FAMILY TREE

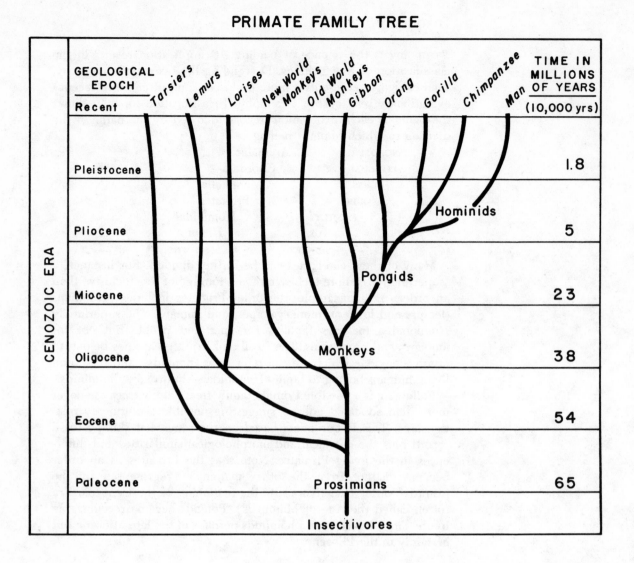

PRE-HOMINIDS

Aegyptopithecus
Dryopithecus
Ramapithecus
Gigantopithecus

Here we are including only those genera that have been suggested as direct ancestors or very close collaterals to the hominids themselves. The reader should be aware that there are many other known fossil primates whose evolutionary relationship is more distant. The very skimpy nature of the fossil evidence for the above groups also should be stressed.

For example, there are no known skull or limb fragments for *Ramapithecus* and *Gigantopithecus*, our knowledge of these forms being entirely restricted to jaws and teeth. Furthermore, *Aegyptopithecus* is known from fragments of fewer than half a dozen individuals, and there is a continuing argument on whether *Ramapithecus* can be distinguished from the range of variation of *Sivapithecus*, a genus that was recognized and named at the end of the first decade of the current century. Even for genus *Dryopithecus*, the best documented of the pre-hominids, the form of the skull is known from only one crushed and distorted specimen, and the form of the post-cranial skeleton is known chiefly from that same individual. Evidently the assessment of ancestral or collateral status of these forms is at best highly speculative and tentative.

DESIGNATION:	*Aegyptopithecus zeuxis*
DATING:	Upper Oligocene, about 28 million years ago.
DISCOVERED BY:	E. L. Simons, American paleontologist (1966 and subsequently).
LOCALE:	Fayum basin, Egypt.
MATERIALS:	Skull, several jaws, teeth, an arm bone, and some foot and tail bones.
SOURCES:	E. L. Simons, 1967, The earliest apes. *Scientific American* 217:28–35 (December).
	J. G. Fleagle, E. L. Simons, and G. C. Conroy, 1975, Ape limb bones from the Oligocene of Egypt. *Science* 189:135–137.
COMMENT:	This may be the best candidate for the common ancestor of the Dryopithecines, and, hence, of modern apes and man. The skull pictured is specimen YPM 23975. The upper incisors are isolated teeth that were not found with the skull. The jaw of the restoration pictured here is a combined cast of two original fragments, YPM 21032 and AMNH 13389.
	The drawing was made from a photograph furnished by Prof. E. L. Simons, Director, Duke University Center for the Study of Primate Biology and History.

Aegyptopithecus zeuxis

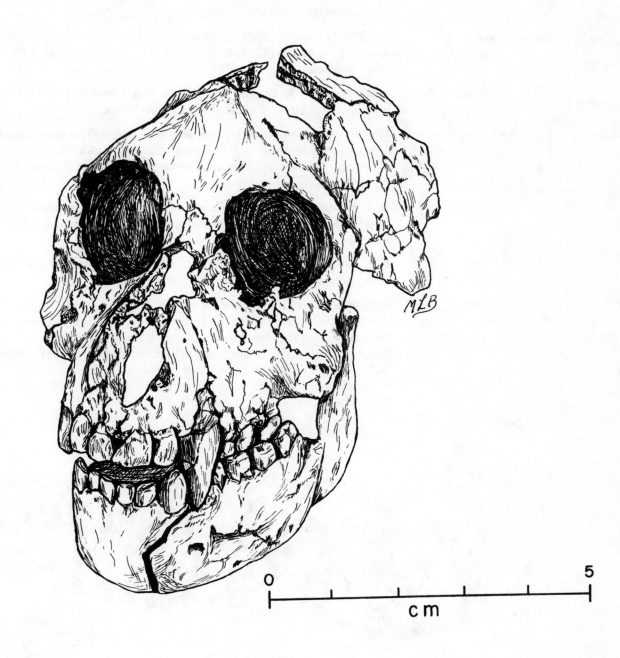

0   5

c m

DESIGNATION:	*Dryopithecus*
	(Members of this genus have also been called "*Proconsul,* (see page 17), *Sivapithecus, Sugrivapithecus,*" and other names by various authors.)
DATING:	Early Miocene in Africa (about 20 million years ago), and Middle Miocene through Middle Pliocene in Eurasia (3 to 16 million years ago).
DISCOVERED BY:	E. Lartet, French prehistorian (1856) and G. E. Lewis, American paleontologist (1910 and subsequently).
LOCALE:	Southern France (1856); Northern India (1910, 1932 and later); Kenya, East Africa (1932, 1948 and later); China (1956 and 1957); and elsewhere.
MATERIALS:	Many jaws and teeth, one skull, a few long bone shafts tentatively attributed, and some foot, ankle, and limb bones.
SOURCES:	William K. Gregory and Milo Hellman, 1926, The dentition of *Dryopithecus* and the origin of man. *Anthropological Papers of the American Museum of Natural History*, Vol. 28, Part 1. New York. pp. 1–123.
	E. L. Simons and D. R. Pilbeam, 1965, Preliminary revision of the Dryopithecinae (Pongidae, Anthropoidea). *Folia Primatologia* 3: 81–152.
	E. L. Simons, 1972, *Primate Evolution: Introduction to Man's Place in Nature.* New York: Macmillan, 322 pp.
COMMENT:	*Dryopithecus* is a geographically widespread genus including species which are ancestral to both modern apes and humans. The diagram shows the well-known *Dryopithecus* or Y-5 pattern of the cusps, typical of the lower molar teeth of fossil and modern anthropoid apes, fossil hominids, and many modern humans.
	Considerably enlarged and drawn after Gregory and Hellman, 1926, Plate X, A.

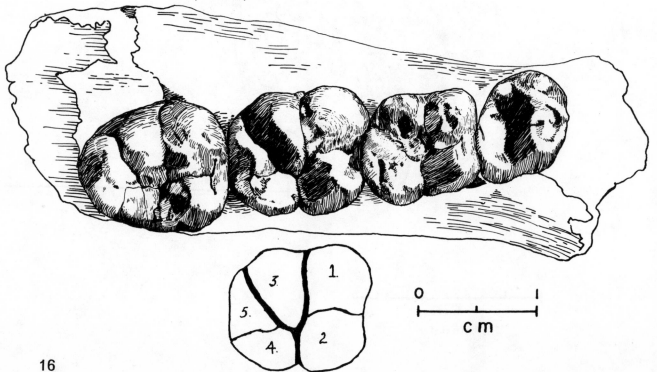

DESIGNATION:	*Dryopithecus africanus* (Known to some as *"Proconsul."*)
DATING:	Early Miocene.
DISCOVERED BY:	Mary D. Leakey, Kenyan prehistorian (1948).
LOCALE:	Rusinga Island (in Lake Victoria), Kenya.
MATERIAL:	Crushed skull, jaws, and teeth.
SOURCE:	W. E. Le Gros Clark and L. S. B. Leakey, 1951, The Miocene Hominoidea of East Africa. *Fossil Mammals of Africa, No. 1.* London: British Museum (Natural History), 117 pp.
COMMENT:	This is the most nearly complete dryopithecine skull discovered thus far. Dr. L. S. B. Leakey and others originally included this specimen in genus *"Proconsul,"* although the recent revision by Simons and Pilbeam (1965) places it in genus *Dryopithecus*. Note that the muzzle is longer and narrower and the incisors smaller than recent apes.

Drawn from a slide taken by W. W. Howells and reproduced with the permission of Professor Howells, Harvard University, and Prof. W. E. Le Gros Clark, Oxford University.

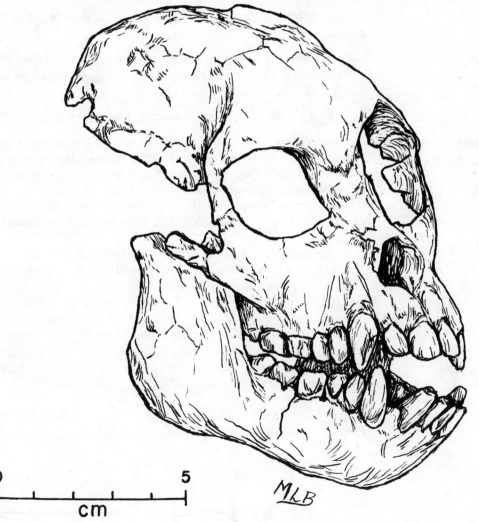

0   5

cm

actual size

DESIGNATION:	*Ramapithecus punjabicus* (Includes part of what has been called *"Kenyapithecus"* in East Africa.)
DATING:	Late Miocene, about 10 to 14 million years ago.
DISCOVERED BY:	G. E. Pilgrim, British paleontologist (1910).
LOCALE:	Northern India (before 1910, 1932), Kenya (1962), and tentatively in Europe (Swabian Alps, Germany) and China (Yunnan, 1956).
MATERIALS:	Jaws and teeth only.
SOURCES:	G. E. Pilgrim, 1915, New Siwalik primates and their bearing on the question of the evolution of man and the Anthropoidea. *Records of the Geological Survey of India*, Vol. 45, Part 1, pp. 2–73. G. E. Lewis, 1934, Preliminary notice of new man-like apes from India. *American Journal of Science* 27:161–179. E. L. Simons, 1961, The phyletic position of *Ramapithecus. Postilla*, Yale University Peabody Museum of Natural History, No. 57:1–9. ————, 1963, A critical reappraisal of Tertiary primates. In J. Buettner-Janusch (ed.), *Evolutionary and Genetic Biology of Primates*, Vol. I. New York: Academic Press, pp. 66–130. ————, 1967, Unravelling the age of earth and man. *Natural History* 76:52–59. ————, 1968, A source for dental comparison of *Ramapithecus* with *Australopithecus* and *Homo. South African Journal of Science* 64:92–112. ————, 1977, Ramapithecus. *Scientific American* 236:28–35 (May).
COMMENT:	Debate continues concerning the correct taxonomic placement of *Ramapithecus*. Many experts feel that it is indeed a member of the general Dryopithecine group; some feel that there is no reason to separate it generically from *Dryopithecus* proper; and there is some recent support for the idea that it is best regarded as belonging to the Dryopithecine genus that Pilgrim identified as *Sivapithecus* in 1910. Other claims have been advanced that *Ramapithecus* is a true hominid; that is, a bipedal, terrestrial, tool-using creature, but most experts are unwilling to push speculation this far. There is general agreement that *Ramapithecus*, a derivative of the Dryopithecine complex, is ancestral to *later* ground-dwelling, bipedal, and tool-using creatures; in other words, ancestral to all true hominids including modern humans. Enough specimens of upper and lower jaws and teeth were consulted to assure that those pictured are reasonably accurate, but the cranial vault must be counted as just an educated guess. The drawing on page 19 is from a half-model made by the Exhibits Department of the University Museum in Ann Arbor, Michigan.

Ramapithecus punjabicus

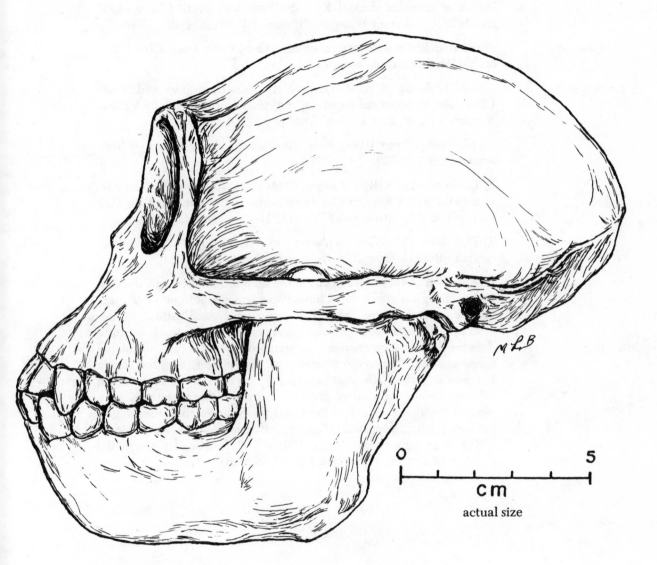

0   5

cm

actual size

DESIGNATION:	*Gigantopithecus* (Two species of this genus have been identified to date: *blacki* from China and *bilaspurensis* from India.)
DATING:	Early Pliocene in India (5 million years or more) and early Middle Pleistocene (tentatively) in China.
DISCOVERED BY:	G. H. R. von Koenigswald, Dutch paleontologist (1935).
LOCALE:	Out of context in a Hong Kong pharmaceutical outlet (1935); from limestone caves in central Kwangsi Province, South China (1955 and 1958); and from Bilaspur, Himachal Pradesh, India (1968).
MATERIALS:	Three mandibles and dozens of individual teeth from China, and one isolated mandible from India.
SOURCES:	Franz Weidenreich, 1945. Giant early man from Java and South China. *Anthropological Papers of the American Museum of Natural History*, Vol. 40, Part 1. New York. pp. 1–134.
	Kwang-chih Chang, 1962. New evidence on fossil man in China. *Science* 136:749–760.
	E. L. Simons and S. R. K. Chopra, 1969. *Gigantopithecus* (Pongidae, Hominoidea): A new species from north India. *Postilla*, Yale University Peabody Museum of Natural History, No. 138:1–18.
	T. D. White, 1975, Geomorphology to paleoecology: *Gigantopithecus* reappraised. *Journal of Human Evolution* 4:219–234.
COMMENT:	The relation of *Gigantopithecus* to the hominid line remains unclear. If the dating is correct, the Chinese specimens would occur too late in time to be anything more than just a large-sized extinct anthropoid ape. The Indian find, however, is early enough to be a candidate for the ancestor of later true hominids, but it is just an isolated specimen and the only primate so far proven for the Early Pliocene anywhere. No post-cranial remains are known, and we still have no idea what kind of creature *Gigantopithecus* actually was, beyond the fact that it had jaws and teeth larger than the biggest modern gorilla. Illustrated is *Gigantopithecus* mandible III (China). The drawing on page 21 is from a cast furnished through the courtesy of Prof. E. L. Simons, Duke University.

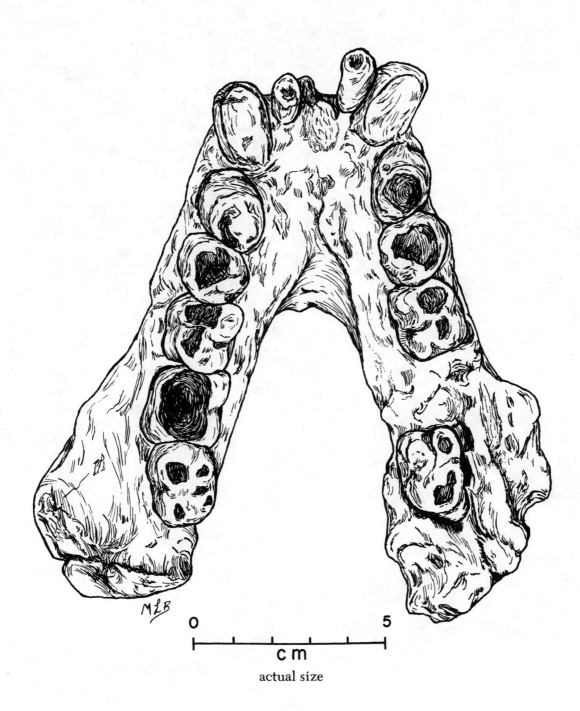

0   5

c m

actual size

HOMINIDS

Australopithecines
Homo erectus
Neanderthals
Moderns

Here we include all those forms which depend for their survival on learned behavior and traditions passed from one generation to the next, that is, on culture. The presence of tools, of course, is a good clue indicating the existence of culture in the larger sense, but for the earliest hominids more than a million years prior to the first evidence for stone tools, we have to make our judgments on the basis of anatomy alone. Their brain size was no larger than that of an ape, but they lacked enlarged and projecting canine teeth, and they walked upright on two feet: i.e., they could neither defend themselves effectively by biting nor could they run rapidly from any quadrupedal predators inclined to prey upon them. We infer from this that, with their hands freed from locomotor duties, they must have wielded implements in their defense. Bipedalism, then, is the indicator that survival is dependent upon culture to the extent that anatomy has become modified as a consequence. Such a creature we term a hominid.

Identification of the various categories of hominid can be an arbitrary matter. For the fossils, the existence of the categories depends as much on the accident of the discovery of major blocks of material as it does on clear-cut morphological features. This, coupled with the fact that the more recent categories evolve out of the earlier ones, means that there is always difficulty in assigning intermediary specimens to either one category or another.

AUSTRALOPITHECINES

In general this category encompasses all those early hominids with small brains and large molars. It includes those forms variously referred to as *"Paranthropus," "Plesianthropus," "Zinjanthropus,"* and probably a number of individuals that have been called *"Homo"* with premature enthusiasm.

There have been a number of serious stumbling blocks in the way of our understanding of the Australopithecines. First there was a problem concerning their dating. The first Australopithecines to be discovered were in South Africa in limestone deposits that could not be directly dated by radiometric means. More recent discoveries of similar forms in datable deposits in East Africa have gone some way toward resolving this problem.

Another problem lay in the nature of our expectations. Single specimens were taken to represent the typical form of whole groups, and other specimens that differed were assumed to belong to different taxonomic categories. It now appears that the average difference in size and robustness between males and females (sexual dimorphism) was greater than has been true for more recent hominids.

We use a single generic designation, *Australopithecus,* for all Australopithecines. Brain size is roughly one third of the modern average and there is more than twice as much chewing surface in the molar teeth. The canine teeth do not project, however, and serve the same effective function as the incisors which are within the modern range of variation. They were bipeds, and even though their bipedalism may have been slightly more efficient than that of more recent hominids, they could not begin to match the speed of quadrupedal predators. Slow of foot and lacking defensively enlarged canines, they must have relied on hand-held weapons for survival purposes.

The earliest Australopithecines are those from the Middle Pliocene at Hadar in Ethiopia and Laetolil in Tanzania some 3.5 million years ago. The specific designation remains disputed although their characteristics are closer to those of accepted specimens of *Australopithecus africanus* than to any other group. Stone tools known as Oldowan tools appear for the first time 1.5 million years later, and, as the Lower Pleistocene gets under way, it is clear that one form of Australopithecine has stressed the development of teeth and chewing muscles to the extent that we recognize it as a separate species, *Australopithecus boisei,* while another has diverged in a cerebral direction and given rise to *Homo erectus.* Just where, when, and how this split occurred is still unclear.

AUSTRALOPITHECINE SITES

1. Taung
2. Kromdraai
3. Sterkfontein
4. Swartkrans
5. Makapan
6. Olduvai Gorge
7. Laetoli
8. Omo (Ethiopia)
9. Koobi Fora (East Turkana, Kenya)
10. Hadar (Ethiopia)

////// Tentative discoveries of Oldowan tools in the Lower Pleistocene.

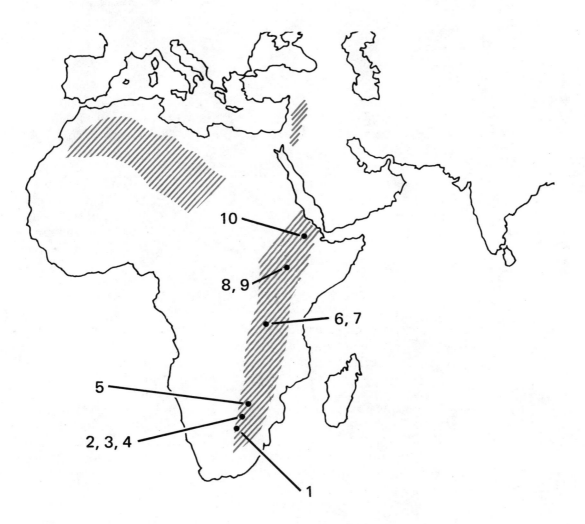

OLDOWAN TOOLS (Pebble Tools):from Bed I of Olduvai Gorge, Tanzania. (Reprinted from K. P. Oakley 1950, *Man The Tool Maker*, 2nd ed., London: British Museum [Natural History], by permission of the Trustees of the British Museum [Natural History].)

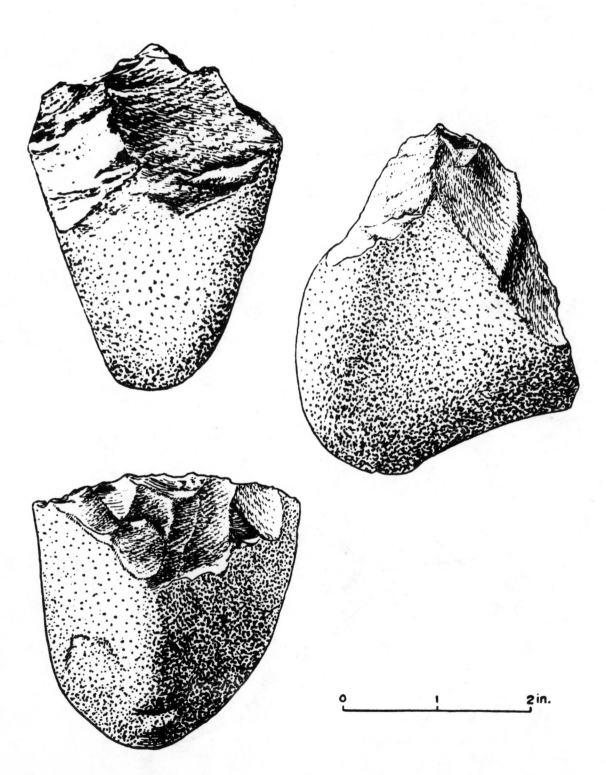

DESIGNATION:	**"Taung"**
	(*Australopithecus africanus.*)
DATING:	Lower Pleistocene.
DISCOVERED BY:	Quarry workers and Prof. R. A. Dart (1924).
LOCALE:	Taung (once incorrectly spelled "Taungs"), Republic of South Africa, southwest of Johannesburg near the edge of what at the time was called British Bechuanaland.
MATERIAL:	Endocranial cast, facial skeleton, jaws, and teeth of a juvenile.
SOURCE:	R. A. Dart, 1925, *Australopithecus africanus:* the man-ape of South Africa. *Nature* 115:195–199.
COMMENT:	Pictured is the immature specimen, equivalent in relative developmental stage to a six-year-old modern child, which first made us aware of an entire stage in human evolution.
	Drawn from a photograph taken in the laboratory of Prof. P. V. Tobias, Department of Anatomy, University of the Witwatersrand, Johannesburg, South Africa. Reproduced with permission.

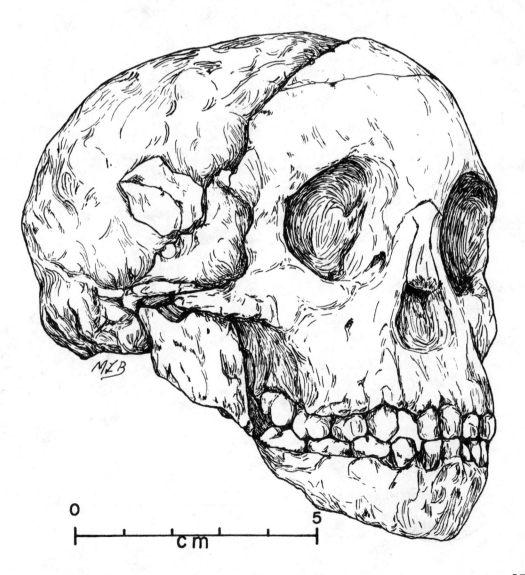

0   5
cm

DESIGNATION: **"Taung"**

COMMENT: This is a side view of the same fossil. The vault of the skull was lost in the quarry blast that brought the specimen to light. What is visible is the endocranial cast; that is, the sediment which filtered into the skull after death and became solidified during fossilization.

Drawn from a photograph taken in the laboratory of Prof. P. V. Tobias, Johannesburg, South Africa.

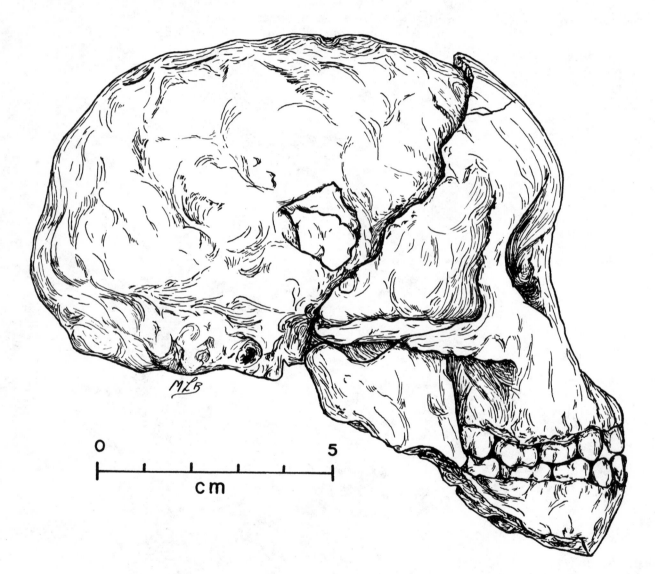

0   5
cm

DESIGNATION: **"Taung"**

COMMENT: This is an occlusal view (considerably enlarged) of the mandible of Dart's find. The teeth at the back of the row are the first permanent molars. In a modern child they would be the "six-year molars." We do not know the age at which they erupted in the Australopithecines. It might have been a little earlier than age six. Both first deciduous molars were slightly damaged when the specimen was chiseled from the rock and should not be regarded as typical of Australopithecine form.

Drawn from a photograph taken in the laboratory of Prof. P. V. Tobias.

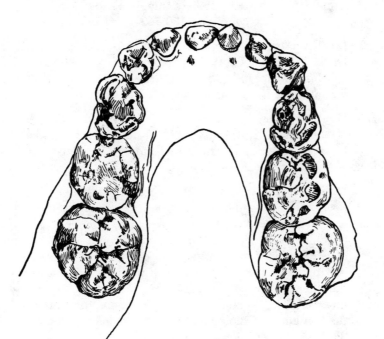

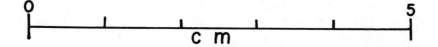

0   5

c m

DESIGNATION:	**"Lucy"** (Afar Locality 288–1, *Australopithecus* cf. *africanus* ?, recently called *Australopithecus afarensis.*)
DATING:	Beginning of the Late Pliocene, 2.9 to 3.0 million years ago using potassium-argon, fission track, and polarity assessment techniques.
DISCOVERED BY:	Dr. Donald C. Johanson, American anthropologist (1974).
LOCALE:	Hadar in the Afar depression 500 kilometers northeast of Addis Ababa, Ethiopia.
MATERIALS:	About 40 percent of an adult female skeleton, including a mandible, some skull fragments, ribs, vertebrae, leg and arm bones, and a pelvis.
SOURCES:	M. Taieb, D. C. Johanson, Y. Coppens, R. Bonnefille, and J. Kalb, 1974, Découverte d'hominidés dans les séries Plio-Pléistocènes d'Hadar (Bassin de l'Awash; Afar, Ethiopie). *Comptes Rendus des Séances de l'Académie des Sciences de Paris* 279:735–738.
	Donald C. Johanson, 1976, Ethiopia yields first "family" of early man. *National Geographic* 150:790–811 (December).
	J. L. Aronson, T. J. Schmitt, R. C. Walter, M. Taieb, J. J. Tiercelin, D. C. Johanson, C. W. Naeser, and A. B. M. Nairn, 1977, New geochronologic and palaeomagnetic data for the hominid-bearing formation of Ethiopia. *Nature* 267:323–327.
	D. C. Johanson, T. D. White, and Y. Coppens, 1978, A new species of the genus *Australopithecus* (Primates: Hominidae) from the Pliocene of eastern Africa. *Kirtlandia, The Cleveland Museum of Natural History* 28:1–14.
COMMENT:	In addition to Lucy, the bones of more than twenty individuals have been found at Hadar, dating to the beginning of the Late Pliocene nearly 3 million years ago. As yet, no complete descriptions are available. This and similar material from Laetoli, 25 miles south of Olduvai in Tanzania, constitute the oldest recognized hominids yet found. In fact, the robust and partially projecting canine teeth, the elongate lower first premolars, and the reinforcements on the inner side of the mandibular symphysis are just half way in between the hominid and the pongid condition—making these a good candidate for the legendary "missing link." But the legs, trunk, and especially the pelvis belonging to Lucy are clearly those of a well-adapted biped, an adult female between 3 and 3 1/2 feet tall, and an unmistakable hominid. Marked differences in size occur between the jaws and teeth of adult specimens, and it seems likely that males were markedly larger than females. Although the specimens from Hadar and Laetoli can be tentatively assigned to *Australopithecus*, and possibly to *A. africanus*, it is certainly the most primitive as well as the earliest hominid material known and suggests that the pongid-hominid split was not much farther back in time. Drawn from a photograph with the permission of Dr. Donald C. Johanson, Cleveland Museum of Natural History.

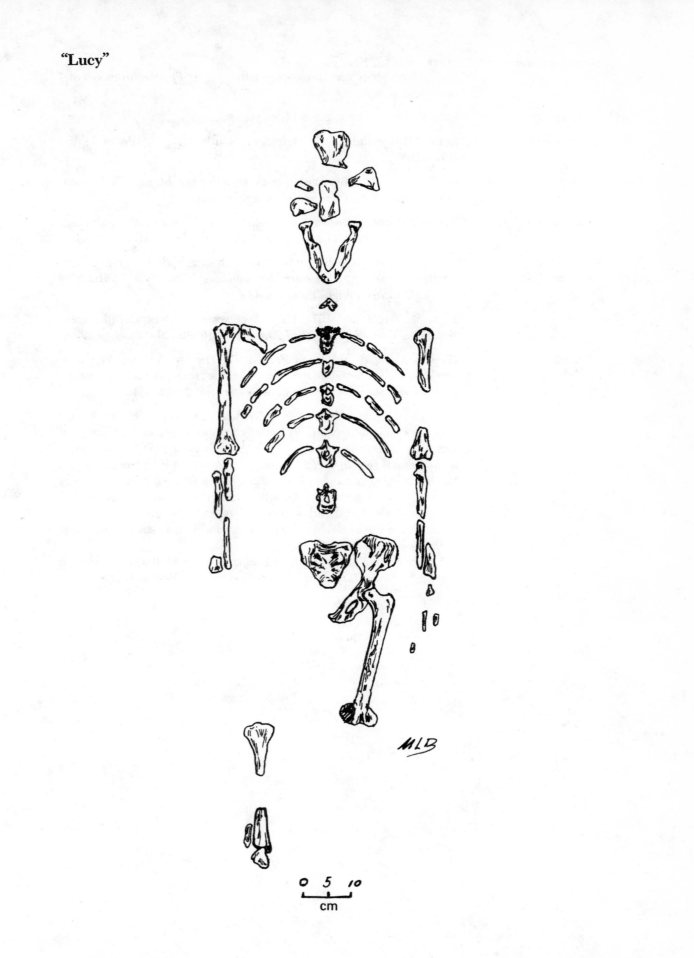

0 5 10
cm

DESIGNATION:	**Sterkfontein, Sts. 14** (*Australopithecus africanus,* originally called *"Plesianthropus trans-vaalensis."*)
DATING:	Late Pliocene, about 2.5 million years ago.
DISCOVERED BY:	Robert Broom and J. T. Robinson, South African paleontologists (1947–1948).
LOCALE:	Sterkfontein, a former lime quarry 40 miles northwest of Johannesburg in the central Transvaal, South Africa.
MATERIALS:	The nearly complete vertebral column, pelvis, some rib fragments, and part of a femur of a very small adult female Australopithecine.
SOURCES:	Milford H. Wolpoff, 1973, Posterior tooth size, body size, and diet in South African gracile Australopithecines. *American Journal of Physical Anthropology* 39:375–394.
	Henry M. McHenry, 1974, How large were the Australopithecines? *American Journal of Physical Anthropology* 40:329–340.
	Charles A. Reed and Dean Falk, 1977, The stature and weight of Sterkfontein 14, a gracile Australopithecine from Transvaal, as determined from the innominate bone. *Fieldiana* 33:423–440.
COMMENT:	The best estimates make this individual just over 4 feet in stature and not much more than 50 pounds in body weight. The range of variation in stature for the South African Australopithecines runs from about 4 feet to 5 1/2 feet. Although Sts. 14 is at the bottom of the South African range of variation, it is slightly larger than Lucy from Hadar in Ethiopia. As with "Lucy," the broad flare of the pelvis and the distance between pelvis and rib cage are far more human than apelike. This already small individual has been further greatly reduced to fit on the page. Drawn from a photograph and reproduced with the permission of Dr. C. K. Brain, Transvaal Museum, Pretoria, South Africa.

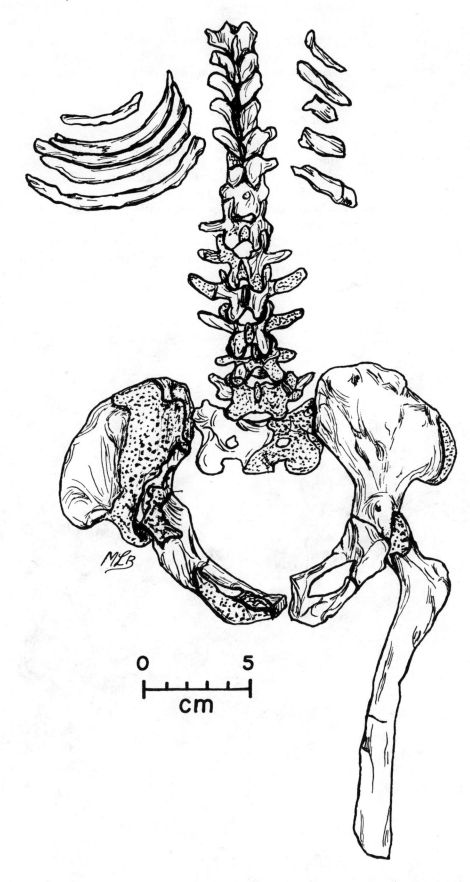

0   5

cm

PELVIC COMPARISONS: AUSTRALOPITHECINE VS. MODERN VS. GORILLA

1. *Australopithecus africanus* from Sterkfontein, Sts. 65.
2. *Australopithecus africanus* from Sterkfontein, Sts. 14s.
3. *Australopithecus africanus* (sometimes called *"robustus"*) from Swartkrans, Sk. 50.
4. *Homo sapiens.*
5. *Gorilla gorilla.*

COMMENT:

The broad flare of the upper part (ilium) of the pelvis both in the Australopithecines and in modern man is a reflection of the erect bipedal mode of locomotion. This is in marked contrast to the gorilla pelvis which is narrower and more elongated in the comparable portion, reflecting its essentially quadrupedal mode of locomotion.

The Australopithecine and modern pelves were drawn from a photograph taken in the Transvaal Museum in Pretoria, South Africa, and are reproduced here with the permission of Dr. C. K. Brain.

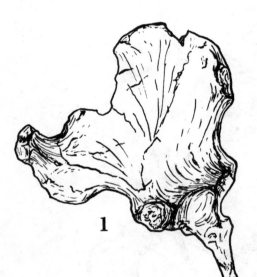

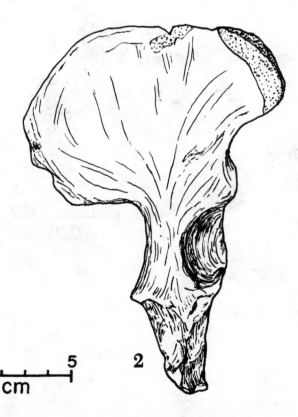

0   5
cm

34

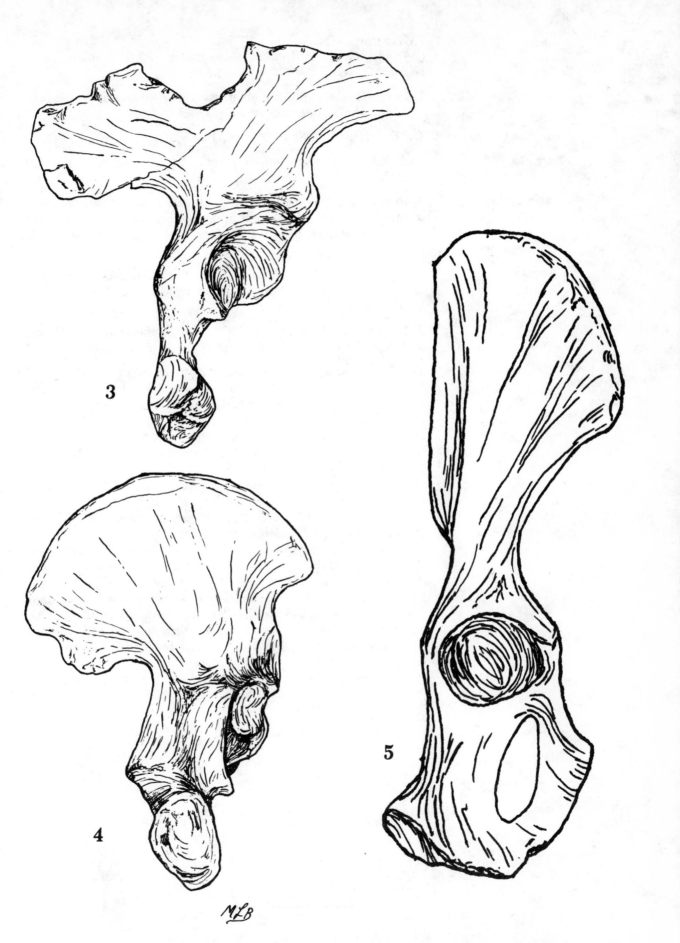

3

4

5

MℱB

35

DESIGNATION:	**"Mrs. Ples," Sts. 5**
	(*Australopithecus africanus*, once called *"Plesianthropus trans-vaalensis."*)
DATING:	Late Pliocene, about 2.5 million years ago by faunal correlations with directly datable sites in East Africa.
DISCOVERED BY:	Dr. Robert Broom, South African Paleontologist (1947).
LOCALE:	Sterkfontein, 40 miles northwest of Johannesburg in the central Transvaal, South Africa.
MATERIALS:	Skull and mandible pictured belong to different individuals. Fragments of skull, mandibles, teeth, and post-cranial pieces of a number of individuals have been found along with Oldowan-type pebble tools.
SOURCES:	Robert Broom, J. T. Robinson, and G. W. H. Schepers, 1950, Sterkfontein ape-man *Plesianthropus. Transvaal Museum Memoirs*, No. 4. Pretoria, South Africa. 104 pp.
	T. D. White and J. M. Harris, 1977, Suid evolution and correlation of African hominid localities. *Science* 198:13–21.
COMMENT:	The skull is catalogued as Sts. 5, or, as Broom called it, "Mrs. Ples." It had been damaged by the quarry blast which led to its discovery, with the top of the skull separated from the lower face-bearing portion. The cap broke off again when it was dropped in the laboratory. After this, it was set back on in a bed of plaster, with the result that the height of the skull appears greater than it should be. The mandible is from specimen Sts. 52b. Note that it came from a much smaller individual than the skull. If the teeth of the skull were restored, the mandible would be forced even lower and the fact that it is too small would be greatly emphasized. The skull was originally classified *"Plesianthropus transvaalensis"* but it is clearly a small- to average-sized member of *Australopithecus africanus*. Associations and dating of the Transvaal deposits are difficult since they apparently accumulated as the result of material falling into limestone sinkholes from the top. The particular species of fossilized pigs in these deposits occur elsewhere in Africa in sites that *can* be directly dated by the potassium-argon technique, and this faunal correlation is what suggests that the date for Sterkfontein must be on the order of 2.5 million years.
	Drawn from a photograph and reproduced with the permission of Dr. C. K. Brain, Transvaal Museum, Pretoria, South Africa.

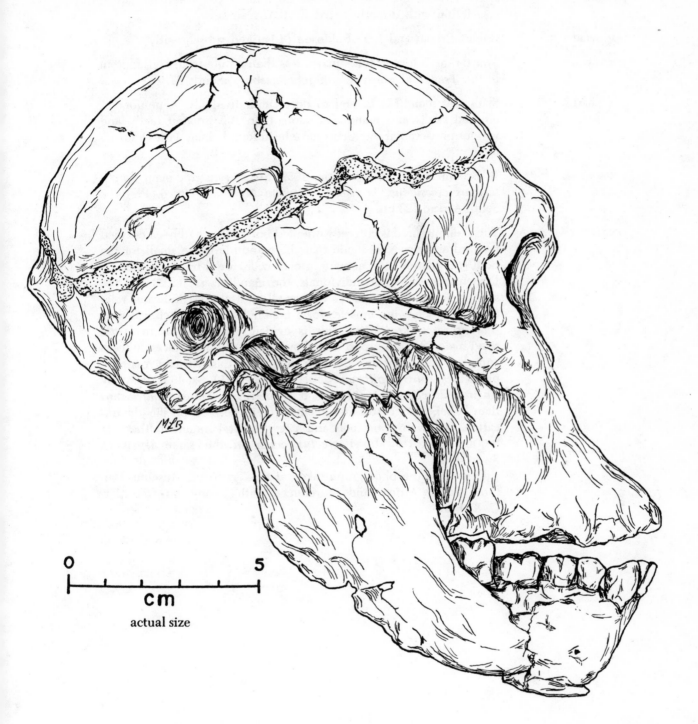

0   5

cm

actual size

DESIGNATION:	**Swartkrans, Sk. 46** (*Australopithecus* cf. *africanus* ?, once called "*Paranthropus crassidens*" and also "*Australopithecus robustus*.")
DATING:	Early Pleistocene, about 1.75 to 1.8 million years ago by faunal correlation with directly dated East African sites.
DISCOVERED BY:	Robert Broom and J. T. Robinson (1949 and subsequently).
LOCALE:	Swartkrans, a former lime quarry less than a mile from Sterkfontein in the Transvaal northwest of Johannesburg, South Africa.
MATERIALS:	Skull and mandible belong to different individuals. A number of crushed skulls and mandibles, more than 100 isolated teeth, and various post-cranial fragments have been found along with Oldowan tools.
SOURCE:	R. Broom and J. T. Robinson, 1952, Swartkrans ape-man, *Paranthropus crassidens. Transvaal Museum Memoirs*, No. 6. Pretoria, South Africa. 123 pp.
COMMENT:	The skull pictured is correctly oriented but its face had been crushed forward and up. Note again that the mandible is too small for the skull (the mandible is actually a perfect fit for "Mrs. Ples," Sts. 5). Drawn from different individuals, the mandible is Sk. 23 and the skull is Sk. 46. The presence of a small sagittal crest indicates relatively large chewing muscles for a hominid. Originally classified as "*Paranthropus crassidens*" this is evidently a medium- to large-sized male (?) of a group that shows some of the features of teeth and skull vault that are found in the robust East African Australopithecines, *Australopithecus boisei* (see page 41). Sk. 46 is definitely *Australopithecus*, but although it is metrically closer to *africanus* than *boisei*, there is still disagreement on its correct specific identification. Oldowan stone tools and a few skeletal fragments that can be attributed to *Homo erectus* occur in the same levels at Swartkrans. Drawn from a photograph taken at the Transvaal Museum, Pretoria, South Africa, and reproduced with permission from Dr. C. K. Brain.

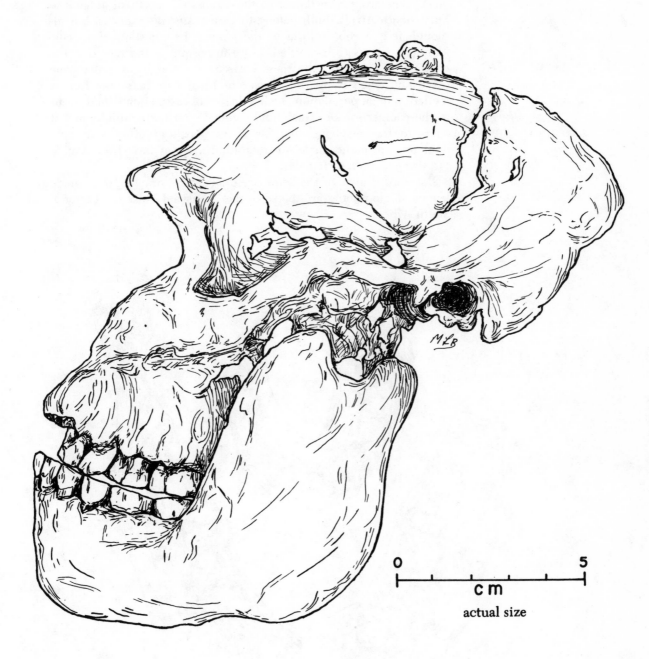

0   5

c m

actual size

DESIGNATION:	**Australopithecine Sexual Dimorphism**
SOURCE:	C. L. Brace, 1973, Sexual dimorphism in human evolution. *Yearbook of Physical Anthropology 1972*, Vol. 16, pp. 31–49.
COMMENT:	Both these mandibles, Sk. 12 (the large one) and Sk. 74 (the small one) come from Swartkrans in the Transvaal, north of Johannesburg, South Africa. Both belong to genus *Australopithecus*, but although it is a robust form of the genus, the question of specific identity remains to be settled. More important is the fact that the contrast in size and robustness between them is greater than one could find when comparing a male and a female mandible from a modern human population. Such a degree of distinction would occur in the comparison of a male and a female gorilla mandible, and it leads to the suspicion that, as in gorillas and other terrestrial nonhuman primates, male Australopithecines may have been double the size and bulk of females.

Both are drawn to the same scale since the drawing was made from a single photograph taken in the Transvaal Museum, Pretoria, South Africa.

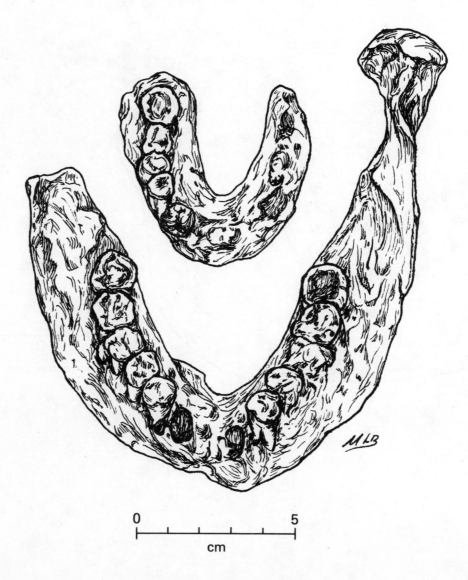

0   5

cm

40

DESIGNATION:	**"Zinj"** (Olduvai Hominid 5, *Australopithecus boisei*, originally called *"Zinjanthropus boisei"* by L. S. B. Leakey, sometimes called *"Australopithecus robustus,"* also referred to as the "Nutcracker Man" and "Dear Boy.")
DATING:	Lower Pleistocene, 1.9 million years ago dated by the potassium-argon dating technique.
DISCOVERED BY:	Mary D. Leakey (1959).
LOCALE:	Olduvai Gorge on the edge of the Serengeti Plain, just northwest of Ngorongoro Crater, Tanzania (Site FLK I, middle of Bed I).
MATERIALS:	Broken skull, upper jaw and teeth, and fragments of lower leg bones, along with Olodwan tools.
SOURCE:	P. V. Tobias, 1967, *The Cranium and Maxillary Dentition of Australopithecus (Zinjanthropus) boisei.* In L. S. B. Leakey (ed.), *Olduvai Gorge,* Vol. 2. Cambridge: Cambridge University Press, 268 pp.
COMMENT:	This is almost certainly a large male specimen of *Australopithecus boisei,* although it was originally called *"Zinjanthropus boisei."* Note the reconstruction between the front and back portions of the skull. Changes in the gap could greatly alter the size of the skull. The rear of the sagittal crest is preserved, but its forward extension is simply an educated guess. As with the Swartkrans skull pictured, it certainly indicates relatively large chewing muscles compared to most other known hominids (for example, the male gorilla). Note, however, the completely hominid canine tooth. The mandible is only a guess. Claims that the Peninj mandible (Lake Natron) is a good fit are not valid since it is substantially too small. Drawn from a photograph taken in Dar-es-Salaam and reproduced with the permission of Dr. L. S. B. Leakey and Prof. P. V. Tobias.

"Zinj"

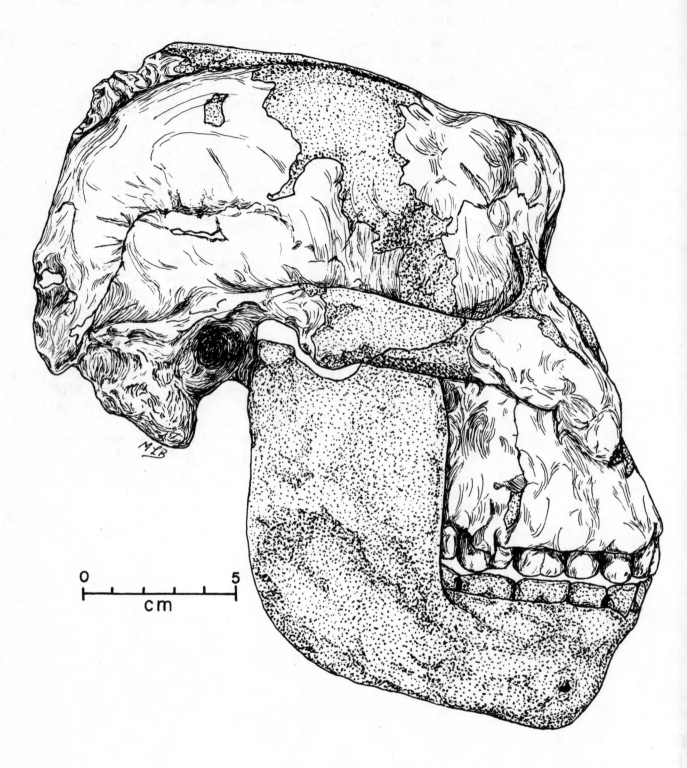

0 ——————— 5
cm

DESIGNATION: **"Zinj"**

COMMENT: Another view of Olduvai H 5 showing the enormous size (but hominid form) of the face in comparison with the relatively small brain case (one-third the size of modern man).

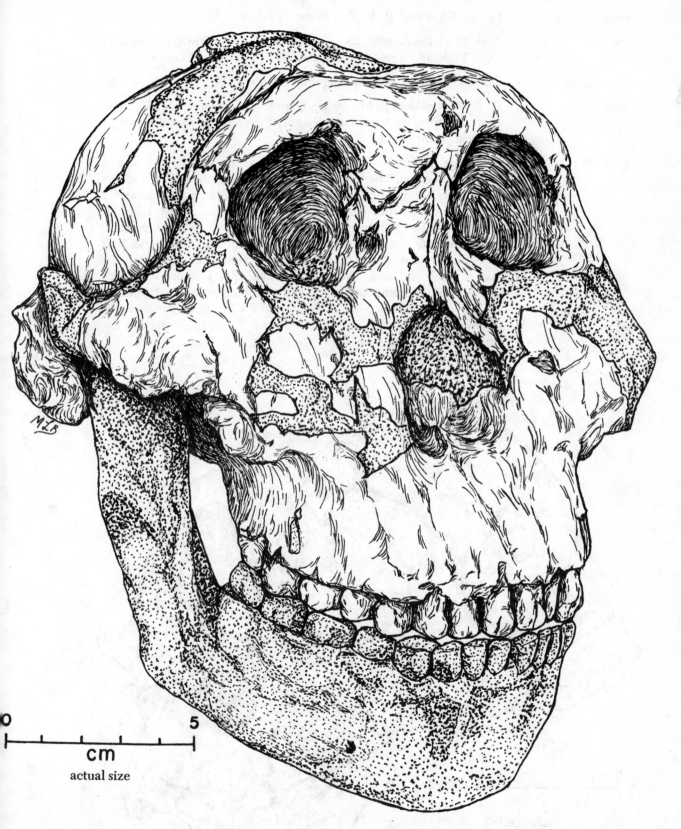

0   5

cm

actual size

43

DESIGNATION:	**ER 406**
	(KNM-ER 406, *Australopithecus boisei.*)
DATING:	Lower Pleistocene, about 1.7 million years ago dated by the potassium-argon technique.
DISCOVERED BY:	Dr. M. Epps and R. E. F. Leakey (1969).
LOCALE:	Area 10 at Ileret, over 25 miles north of Koobi Fora, just east of Lake Turkana (formerly Lake Rudolf) in northeast Kenya.
MATERIALS:	One complete cranium lacking only the teeth.
SOURCE:	R. E. F. Leakey, J. M. Mungai, and A. C. Walker, 1971, New Australopithecines from East Rudolf, Kenya. *American Journal of Physical Anthropology* 35:175–186.
COMMENT:	Brain size is just over 500 cubic centimeters which is typically Australopithecine. The face is greatly expanded, the palate and the tooth roots are very large, and the extensive attachment areas for the muscles of mastication are characteristic for the robust Australopithecines. The well-developed cheek bones and broad zygomatic arches indicate great masseter muscle development, and the sagittal crest shows that the temporal muscles completely encased the cranial vault just as is the case in "Zinj." ER 406, properly considered *Australopithecus boisei*, was a contemporary neighbor of the fully *erectus* ER 3733 (see page 67) indicating that two distinct kinds of hominid must have coexisted during the lower Pleistocene in East Africa.
	Drawn from a cast.

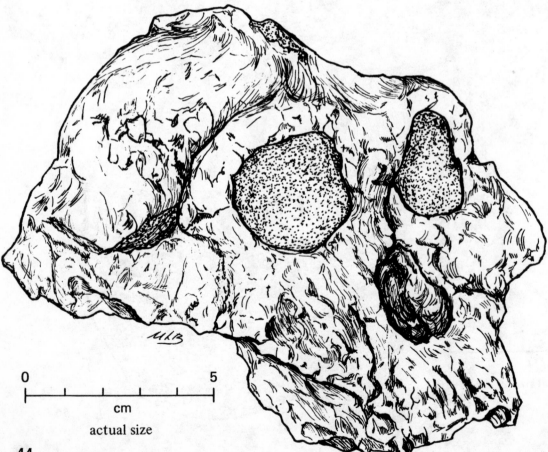

0   5

cm

actual size

DESIGNATION:	**"Pre-Zinj"**
	(Olduvai Hominid 7, *Australopithecus africanus*, also referred to as the "Pre-Zinj juvenile" or "adolescent" and *"Homo habilis.")*
DATING:	Lower Pleistocene, 1.9 million years ago.
DISCOVERED BY:	Dr. L. S. B. Leakey, Kenyan prehistorian (1960).
LOCALE:	Olduvai Gorge, Tanzania (Site FLKNN I, middle of Bed I).
MATERIALS:	An adolescent mandible found with a fragmentary occipital and two broken parietal bones, a clavicle, hand bones, and a nearly complete adult foot.
SOURCES:	See the papers in "The Origin of Man" symposium published in *Current Anthropology* 6:342–431 (October, 1965).
	C. L. Brace, P. E. Mahler, and R. B. Rosen, 1973, Tooth measurements and the rejection of the taxon *"Homo habilis." Yearbook of Physical Anthropology* 1972, Vol. 16, pp. 50–68.
COMMENT:	These remains form the type collection for the claimed taxon *"Homo habilis,"* but since the teeth are indistinguishable from Australopithecine teeth, they are best regarded as belonging to another individual of *Australopithecus africanus.*
	Drawn from a photograph and reproduced with the permission of Dr. L. S. B. Leakey and Prof. P. V. Tobias.

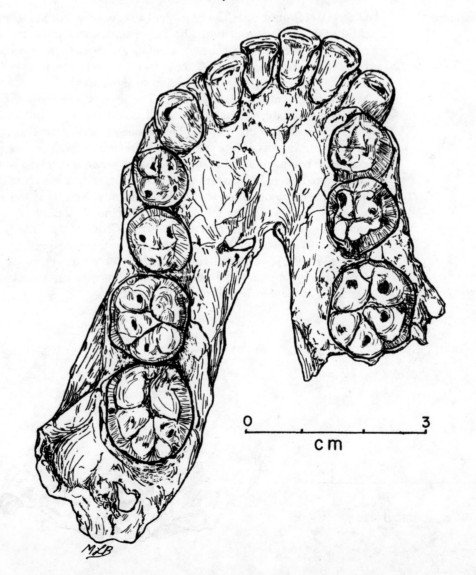

cm

DESIGNATION:	**Olduvai Hominid 16 (OH 16)** (*Australopithecus africanus,* also referred to as "Poor George," "Olduvai George," "Maiko Gully," "*Homo habilis,*" and "*Homo erectus*" by various authorities. It is pictured here with "Chellean Man" (Olduvai Hominid 9) which is generally accepted as a *Homo erectus* skull.)
DATING:	Lower Pleistocene, about 1.7 million years ago for OH 16, and late Lower Pleistocene, about 1.2 million years ago for Olduvai Hominid 9 (OH 9).
DISCOVERED BY:	L. S. B. Leakey (1963 and 1960 respectively).
LOCALE:	Olduvai Gorge, Tanzania (Site FLK II, Maiko Gully, lower Bed II; and site LLK II, Upper Bed II respectively).
MATERIALS:	Fragments of an adult skull and both upper and lower teeth for "George" and one reconstructible skull minus the face for Olduvai Hominid 9.
SOURCES:	L. S. B. Leakey and Mary D. Leakey, 1964, Recent discoveries of fossil Hominids in Tanganyika: at Olduvai and near Lake Natron. *Nature* 202:5–7. Melvin M. Payne, 1966, Preserving the treasures of Olduvai Gorge. *National Geographic* 130:700–709 (November).
COMMENT:	Olduvai Hominid 16, "George," had apparently been a complete skull with the face and both the upper and lower jaws. It washed out of lower Bed II in a heavy rainfall and was then trampled on by a herd of Masai cattle. The facial skeleton was completely demolished leaving only a collection of loose teeth and some badly broken fragments of cranial vault. The reconstruction is therefore only approximate, although the skull could certainly be no larger than it is scaled in the picture. This is one of several finds that have been referred to as *Homo habilis* without benefit of definitive study, adequate comparison, or even, in this case, measurements. The teeth, which are the only measurable parts preserved, are indistinguishable from those of *Australopithecus africanus.* Note that the reconstructed skull, as befits an Australopithecine, is definitely smaller than that of a full-fledged *Homo erectus,* Olduvai Hominid 9. Both are drawn to the same scale since the drawing was made from a single photograph taken in the Centre for Prehistory and Paleontology in Nairobi, Kenya. Printed here with the permission of Dr. L. S. B. Leakey and Prof. P. V. Tobias.

OH 16

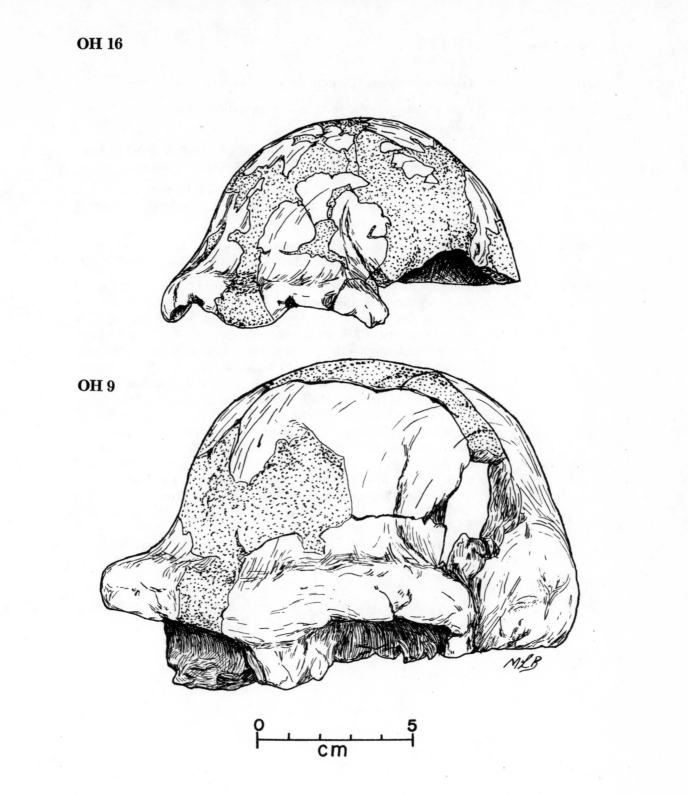

OH 9

47

DESIGNATION:	**ER 1805** (*Australopithecus* cf. *africanus* ?)
DATING:	Lower Pleistocene, 1.7 million years ago.
DISCOVERED BY:	P. Abell, a member of the research team directed by R. E. F. Leakey (1973).
LOCALE:	In the Upper Member of the Koobi Fora Formation at Koobi Fora, area 130 upper, east of Lake Turkana in northeast Kenya.
MATERIALS:	A relatively complete brain case (assembled from fragments), a maxilla with most of the upper teeth, and the body of a mandible with two molars (not pictured here).
SOURCE:	R. E. F. Leakey, 1974, Further evidence of Lower Pleistocene hominids from East Rudolf, North Kenya, 1973. *Nature* 248:653–656.
COMMENT:	The face, palate and teeth are in the size range of the South African Australopithecines; the tooth measurements, in fact, fall right in between the averages for Sterkfontein and Swartkrans. With a cranial capacity of 582 cubic centimeters, the skull was at the large end of the Australopithecine range of variation. Even so, the area for temporal muscle attachment rises so far on each side of the skull that the low parasagittal crests almost touch at the midline. At both sides of the back of the skull, the temporal muscle attachment area meets the neck muscle attachment area and a compound temporo-nuchal crest is formed. This feature is characteristically found in *Australopithecus boisei;* however, in ER 1805 the teeth are too small to assign it with that taxon. On the other hand, the face and teeth are too large and brain too small to warrant inclusion in the genus *Homo*. Yet it was a contemporary of both *Homo erectus* (ER 3733, see p. 66) and *A. boisei* (ER 406, see p. 44) in the East Turkana area. Drawn from a cast.

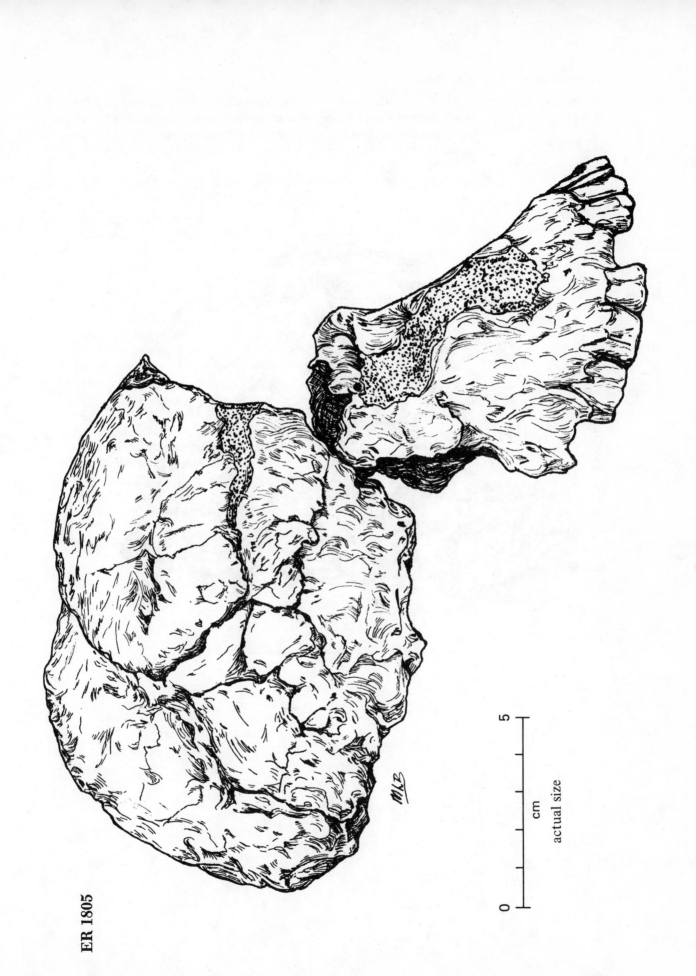

ER 1805

5

cm

actual size

0

49

DESIGNATION: **ER 1805**

COMMENT: Viewed from the rear, it looks especially Australopithecine. The small sagittal crest can be seen at the top, and the compound temporonuchal crest leading to the expanded mastoid area can be seen at the lower left (it is broken off from the right-hand side).

Drawn from a cast.

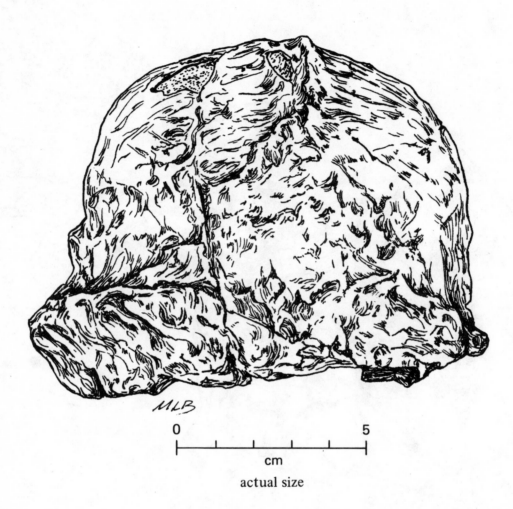

cm

actual size

50

DESIGNATION:	**ER 1813** (*Australopithecus africanus* ?)
DATING:	Lower Pleistocene, 1.7 million years ago.
DISCOVERED BY:	Kamoya Kimeu, a member of the research team directed by R. E. F. Leakey (1973).
LOCALE:	In the Upper Member (provisional) of the Koobi Fora Formation at Koobi Fora, area 123, east of Lake Turkana in northeast Kenya.
MATERIALS:	A fragmented but reconstructible cranium including the face and many of the teeth.
SOURCE:	M. H. Day, R. E. F. Leakey, A. C. Walker and B. A. Wood, 1976, New hominids from East Turkana, Kenya. *American Journal of Physical Anthropology* 45:369–436.
COMMENT:	This is the best candidate available in East Africa for a representative of what is accepted as *Australopithecus africanus* in South Africa. Cranial capacity is 510 cubic centimeters which is average for *A. africanus* and, although the facial skeleton appears on the small side, the teeth also are close to the South African Australopithecine average. It was a contemporary of both *Homo erectus* (ER 3733 see page 65) and *Australopithecus boisei* (ER 406, see page 44) in the East Turkana area. Drawn from a cast.

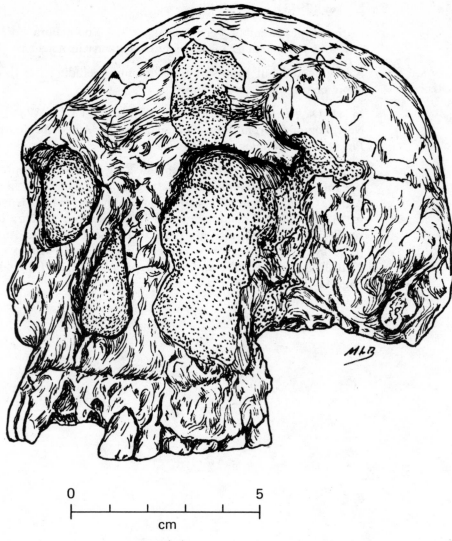

0   5

cm

actual size

DESIGNATION:	**ER 1470** (KNM-ER 1470, *"Homo"* sp. indet., also known as "Homo habilis").
DATING:	Early Lower Pleistocene, just older than the 1.8 million year date of the overlying KBS tuff dated by the potassium-argon technique.
DISCOVERED BY:	Bernard Ngeneo, a member of the research team directed by Richard E. F. Leakey (1972).
LOCALE:	Area 131, about thirteen miles northeast of the camp at Koobi Fora, east of Lake Turkana (formerly Lake Rudolf) in northern Kenya.
MATERIALS:	Relatively complete cranial vault reconstructed from more than a hundred pieces, plus fragments of the facial skeleton including the tooth-bearing part of the maxilla but not the teeth.
SOURCES:	R. E. F. Leakey, 1973, Evidence for an advanced Plio-Pleistocene hominid from East Rudolf, Kenya. *Nature* 242:447–450. ————, 1973, Skull 1470. *National Geographic* 143:818–829 (June).
COMMENT:	The skull was found at the end of August 1972, and Richard Leakey was able to show it to his father, Louis Leakey, who died just over a month later in London. Louis felt that Richard had at last found the early "true *Homo*" that had been the object of his life-long quest. Brain size, at 750 cubic centimeters, is at the bottom end of the range of variation for anything accepted as *Homo,* but distinctly larger than that of any known Australopithecine. The brow ridge enlargement and vault thickening of the later *erectus* representatives are not yet evident. From the size of the palate and the expansion of the area allotted to molar roots, it would appear that ER 1470 retained a fully *Australopithecus*-sized face and dentition. From the combination of cranial and facial features, it would appear that ER 1470 represented a stage intermediate between the Australopithecines and full *Homo erectus* status. It does not fully fit in either category but may represent an example of how the transition took place. Drawn from a cast.

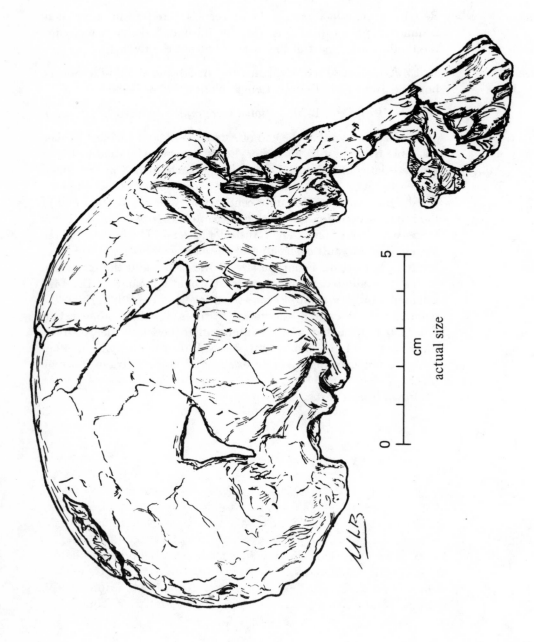

ER 1470

5

cm

actual size

0

54

DESIGNATION:

Palates

Divergent trends of hominid dental evolution as represented by the maxillary (upper) dental arches of an early *Homo sapiens* (top drawing), a robust Australopithecine (bottom), and the condition from which both arose, *Australopithecus africanus,* represented in the middle.

COMMENT:

The palate in the middle belongs to Olduvai Hominid 16 ("George"). The measurements of the individual teeth are those of an average specimen of *Australopithecus africanus* from South Africa and very little different from those of the Hadar and Laetoli Australopithecines of over a million years earlier. However, at the end of the Pliocene or the beginning of the Pleistocene approximately 2 million years ago, the size and form of the ancestral hominid dentition was very much like that pictured in the middle drawing. From that condition, the hominid lineage split with one branch, symbolized by the "Zinj" dentition at the bottom, stressing an enlargement of the molars and an enlargement and "molarization" of the premolars to produce teeth of gorilloid size. This branch, properly labeled *Australopithecus boisei,* became extinct before the end of the Lower Pleistocene. The other branch continued through the Middle and Upper Pleistocene and survives today as modern *Homo sapiens.* The early Upper Pleistocene condition of this branch is illustrated by the palate and dentition of the top drawing, the classic Neanderthal from Le Moustier. Note the trend toward reduction in the molar teeth. The Neanderthal molars are just within the upper limits of the modern range of variation; some of the molar dimensions of the ancestral Australopithecine condition, particularly toward the rear, are larger than those of any more recent human teeth, and all dimensions of the robust Australopithecine (*boisei*) molars are larger than those of more recent hominids. For the incisors, the trend has nearly been the reverse. Starting with small incisors at the Australopithecine stage, size increases to a maximum among the Neanderthals following which there has been a sharp reduction in the Late Pleistocene and post-Pleistocene so that the incisors of modern man are back at Australopithecine size levels.

The *boisei* palate was drawn from a photograph taken in the National Museum of Tanzania, Dar-es-Salaam. The teeth of Olduvai Hominid 16 were set in clay to hold them in place and photographed in the Centre for Prehistory and Palaeontology, National Museum, Nairobi, Kenya. The Le Moustier palate was drawn after the plate in Gregory and Hellman 1926. Casts of both the *boisei* and Le Moustier palates are present in the laboratory in Nairobi, and were photographed with the original "George" dentition. This photograph was used to ensure that the drawings made from the original photographs of the *boisei* and Le Moustier dentitions were exactly to the same scale as that of the Olduvai Hominid 16 dentition. Printed here with the permission of Dr. L. S. B. Leakey and Prof. P. V. Tobias.

Palates

Early *Homo sapiens* (Le Moustier)

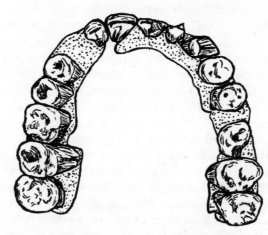

Australopithecus africanus (OH 16)

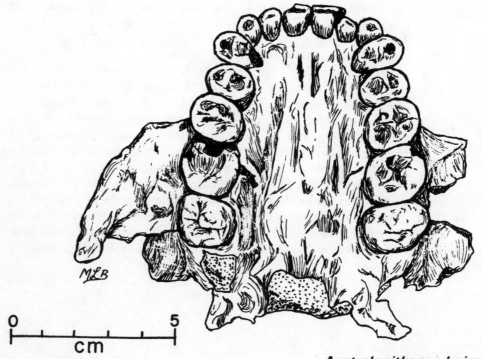

0 ———— 5
cm

Australopithecus boisei (OH 5)

56

HOMO ERECTUS

Fossils assigned to this taxon are also known as "Pithecanthropines" after the original type fossil from Java, *"Pithecanthropus" erectus.*

On the average, brain size was double that of the Australopithecines and two-thirds that of the modern norm. The molars are markedly reduced in comparison with the Australopithecines and fall within the modern range of variation, although right at the upper end. Braincase reinforcements and brow ridges are very robust. Cultural material, when present, is of Lower Palaeolithic form including flakes and/or hand axes. From the nature and quantity of butchered animal remains associated with the stone tools, it is evident that, for the first time, systematic hunting of large game had become a major part of the hominid way of life.

ERECTUS SITES

(Restricted to the most important and best preserved specimens).

1. Java
2. Choukoutien
3. Lan-t'ien (China)
4. Heidelberg
5. Vertesszöllös (Hungary)
6. Arago (France)
7. Ternifine (Algeria)
8. Rabat (Morocco)
9. Koobi Fora (Kenya)
10. Olduvai Gorge
11. Swartkrans
12. Saldanha

Tentative distribution of Lower Palaeolithic tools associated with *Homo erectus.*

LOWER PALAEOLITHIC TOOLS

ACHEULIAN TOOLS: *a*) Lava biface (hand axe) from Ol Orgesailie, Kenya; *b*) Ovate biface from St. Acheul, northwestern France; *c*) Ovate biface from Israel; *d*) Micoquian biface from Suffolk, England.

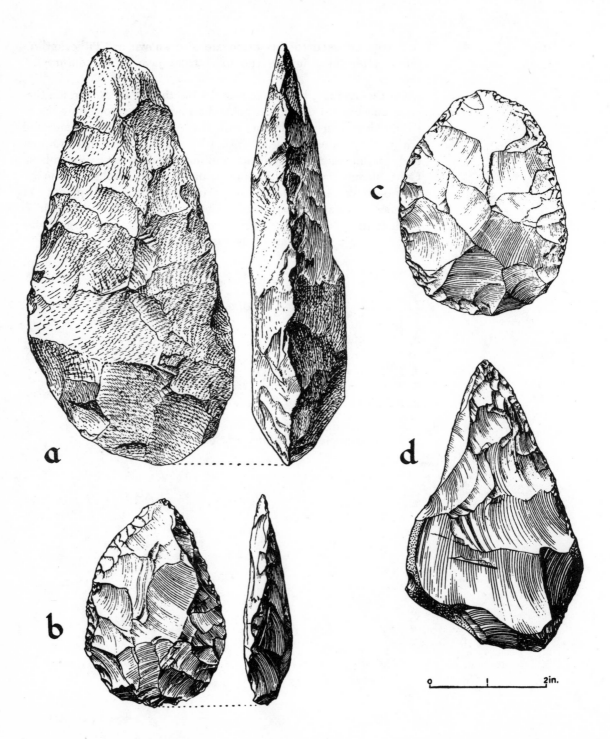

DESIGNATION:	**"Java Man"**
	(*Homo erectus*, formerly referred to as *"Pithecanthropus erectus"* and *"Trinil"*).
DATING:	Middle Pleistocene, about 500,000 years ago.
DISCOVERED BY:	Eugene Dubois, Dutch anatomist (1891).
LOCALE:	Trinil in Central Java.
MATERIALS:	Skull cap, a complete femur (thigh bone), three teeth, and a jaw fragment (actually found in 1890 but not recognized until many years later).
SOURCES:	G. H. R. von Koenigswald, 1949, The discovery of early man in Java and southern China. In W. W. Howells (ed.), *Early Man in the Far East: Studies in Physical Anthropology, No. 1.* Detroit: The American Association of Physical Anthropologists, pp. 83–98.
	————, 1956, *Meeting Prehistoric Man*, translated by Michael Bullock, London: Thames and Hudson, 216 pp.
	Gert-Jan Bartstra, 1976, Contributions to the study of the Palaeolithic Patjitan culture, Java, Indonesia. In J. E. van Lohuizen-de Leeuw (ed.), *Studies in South Asian Culture, Vol. IV*, Institute of South Asian Archaeology, University of Amsterdam. Leiden: E. J. Brill, 121 pp.
COMMENT:	This specimen was formerly called *"Pithecanthropus erectus"* and was the first find of what we now recognize as the *Homo erectus* stage of human evolution. It is now generally agreed that this stage is not generically distinct from modern man, that is, genus *Homo*. Brain size averages around 1,000 cubic centimeters, approximately double that of the Australopithecines and two-thirds that of the modern average. Note the presence of heavy supraorbital ridges. The molar teeth are markedly reduced from Australopithecine size. Lower Palaeolithic types of tools including bifaces ("hand axes") have subsequently been found in some number in Java and on mainland southeast Asia. Evidently the cultural adaptation was the same as that seen in Europe, Africa, and India at the same time.

Drawn after Day, 1965, and reproduced by permission of The World Publishing Company, Cleveland, Ohio, from *Guide to Fossil Man* by Michael Day, Copyright © 1965 by Michael Day.

"Java Man"

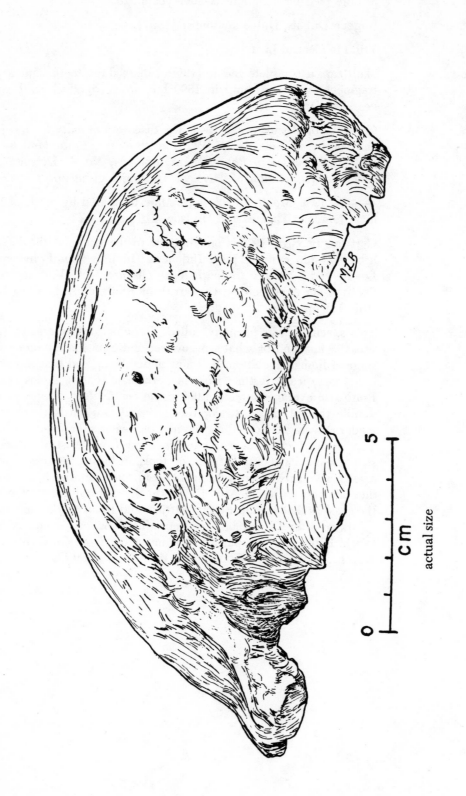

0   cm   5

actual size

DESIGNATION:	**"Pekin Man"** (*Homo erectus,* originally called *"Sinanthropus pekinensis"*).
DATING:	Middle Pleistocene, about 300,000 years ago.
DISCOVERED BY:	W. C. Pei, Chinese paleontologist (1929 and subsequently, although teeth had been found there two years before).
LOCALE:	Choukoutien, 30 miles southwest of Peking, China.
MATERIALS:	A dozen incomplete but identifiable skulls, mandibles, 147 teeth, and some fragments (none complete) of the post-cranial skeleton.
SOURCE:	Franz Weidenreich, 1943, The skull of *"Sinanthropus pekinensis."* *Palaeontologia Sinica,* Vol. 10, Whole Series No. 127, pp. 1–485.
COMMENT:	This population was first called *"Sinanthropus pekinensis";* however, as the principal describer Franz Weidenreich realized, the designation *Homo erectus* is more accurate. The specimen pictured on the next three pages is the most complete representation available for a *Homo erectus* individual but the reconstruction may be slightly misleading. By chance, the most complete Choukoutien skull (the basis for this reconstruction) was that of a small, comparatively lightly built female. The average member of the population was considerably more rugged. Brow ridges formed a single continuous supraorbital torus, and neck muscle attachments for the males were much more prominent. The midportion of the face also is sheer guesswork and results in an unnecessarily muzzlelike appearance for the lower face. The whole collection on which Weidenreich's study was based disappeared when the Japanese invaded China during the second World War and has never reappeared. Drawn from a cast representing a composite reconstruction made by the late Dr. Weidenreich.

"Pekin Man"

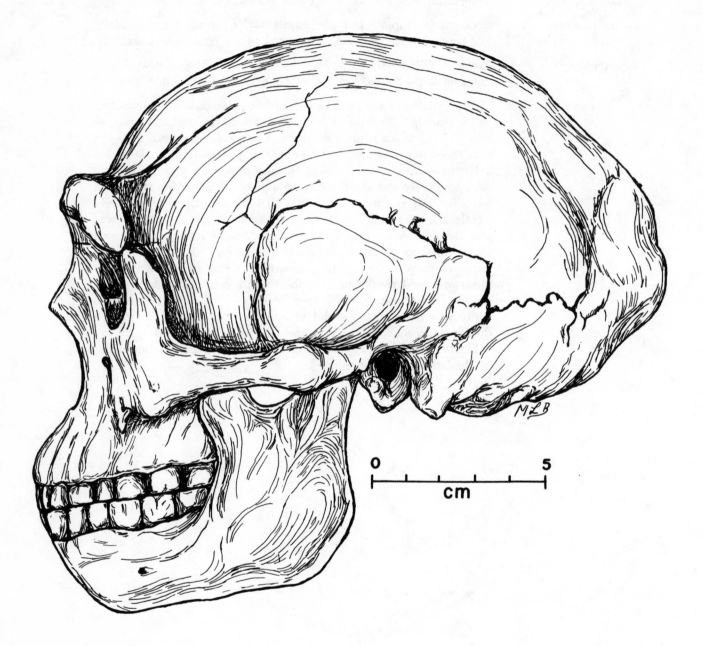

"Pekin Man"

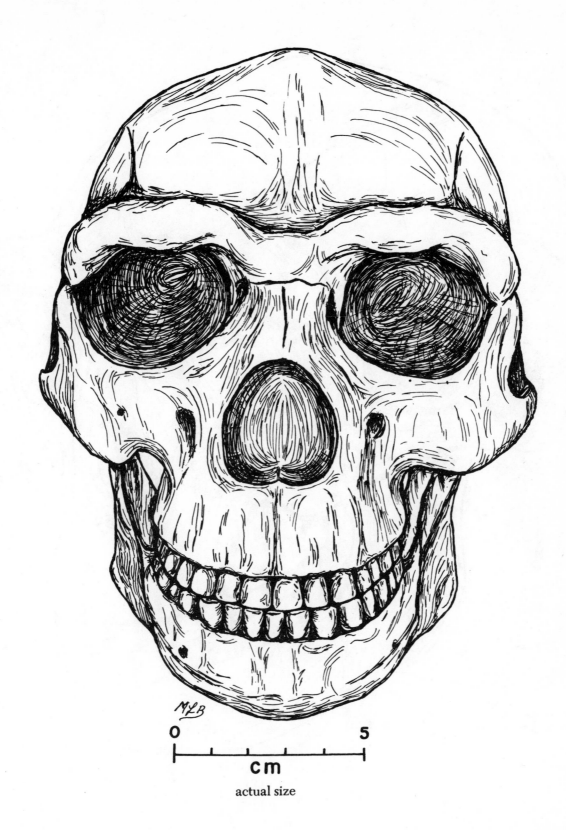

0

5

cm

actual size

"Pekin Man"

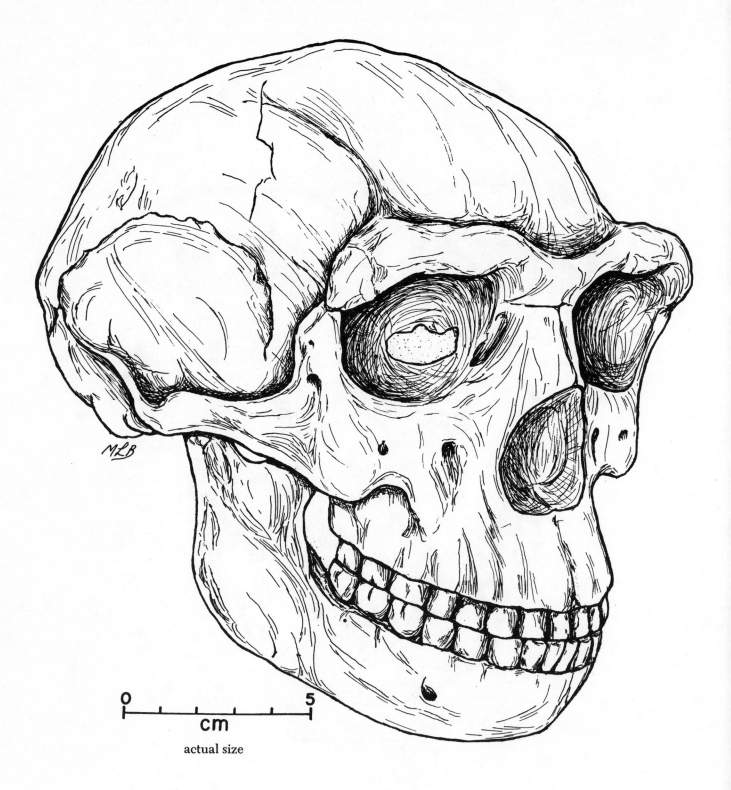

0   5
cm
actual size

DESIGNATION:	**ER 3733** (KNM-ER 3733, *Homo erectus*, "wolf")
DATING:	Lower Pleistocene, 1.7 million years ago. The date is the same as that for ER 406, ER 1805, and ER 1813.
DISCOVERED BY:	Bernard Ngeneo, a member of the research team led by Richard E. F. Leakey (1975).
LOCALE:	Koobi Fora area 104, east of Lake Turkana (formerly Lake Rudolf) in northeast Kenya. The discovery was made in the Upper Member of the Koobi Fora Formation above the KBS tuff but below the Koobi Fora tuff.
MATERIALS:	Relatively complete cranial vault with facial skeleton, palate, and some premolars and molars.
SOURCES:	R. E. F. Leakey and A. C. Walker, 1976, *Australopithecus, Homo erectus,* and the single-species hypothesis. *Nature* 261:572–574. R. E. F. Leakey, 1976, Hominids in Africa. *American Scientist* 64: 174–178. ———, 1976, New hominid fossils from the Koobi Fora Formation in northern Kenya. *Nature* 261:574–576.
COMMENT:	For years, the late Louis S. B. Leakey had been offering now one find now another as evidence for the existence of "true man" back at a time when *Australopithecus* was in existence. Presumably this would deny *Australopithecus* place in the ancestry of *Homo*. For years, each specimen offered as ancient "true man" turned out either not to be ancient, or not to be "true man," and many anthropologists put Dr. Leakey in the category with the "little boy who cried wolf." But just as the little boy eventually did encounter a genuine wolf, so the efforts of the Leakey family eventually did lead to the discovery of a genuine representative of the genus *Homo* that *was* a contemporary of a representative of the genus *Australopithecus*. Both had evidently descended in diverging lines that can be traced to an earlier form identifiable as *Australopithecus africanus*. Drawn from a photograph furnished by the National Museums of Kenya.

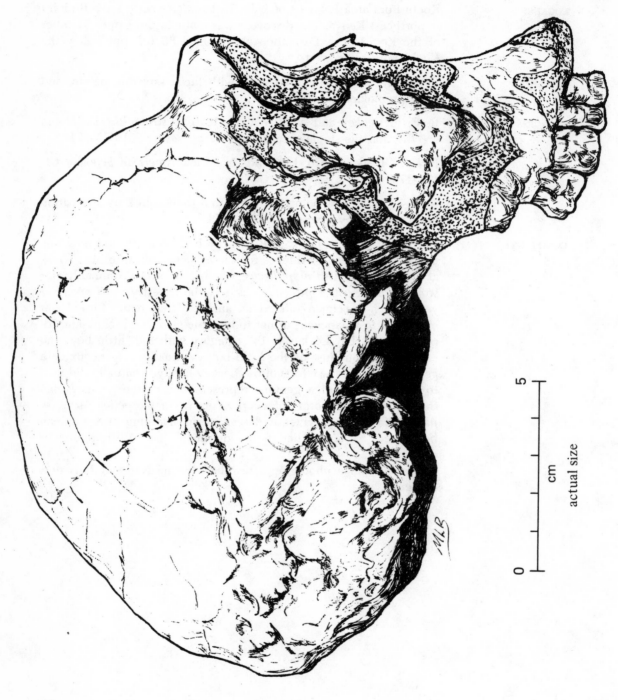

ER 3733

66

DESIGNATION: ER 3733

COMMENT: Brain size, at just over 800 cubic centimeters, is within the lower limits of the *erectus* range of variation. Cranial reinforcements, as exemplified by the thickened brow ridge, are also of *erectus* rather than Australopithecine form, and the face is reduced to the level where hominid face form was to remain without change until the onset of the last glaciation less than 100,000 years ago. There can be no doubt that this is a good representative of *Homo erectus*, one of the earliest and most complete specimens yet found.

Drawn from a photograph furnished by the National Museums of Kenya.

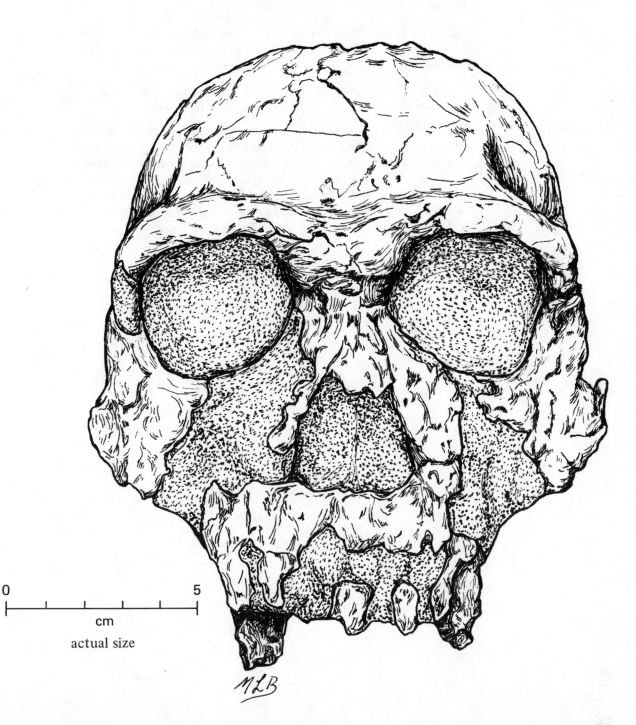

0   5

cm

actual size

| DESIGNATION: | **"Chellean Man"** |
| | Olduvai Hominid 9 (*Homo erectus*, erroneously called *"Homo leakeyi"*). |

DATING: Late Lower Pleistocene, about 1.3 million years ago.

DISCOVERED BY: Dr. L. S. B. Leakey (1960).

LOCALE: Olduvai Gorge, Tanzania (Site LLK II, Upper Bed II).

MATERIALS: One reconstructible skull minus the face. Associated with Lower Palaeolithic stone tools clearly advanced over those in Bed I.

SOURCES: L. S. B. Leakey, 1961, New finds at Olduvai Gorge. *Nature* 189: 649–650.

Gerhard Heberer, 1963, Über einen neuen archanthropinen Typus aus der Oldoway-Schlucht. *Zeitschrift für Morphologie und Anthropologie* 53:171–177.

P. V. Tobias, 1965, New discoveries in Tanganyika: Their bearing on hominid evolution. *Current Anthropology* 6:391–411.

COMMENT: With a capacity of 1000 cubic centimeters and a brow ridge of monumental proportions, this is accepted by most anthropologists as an undisputed representative of *Homo erectus*, even though there has as yet been no definitive study made. It would appear to be a male counterpart of the female Olduvai Hominid 13 (see page 70).

Drawn from a photograph and printed with the permission of Dr. L. S. B. Leakey and Prof. P. V. Tobias.

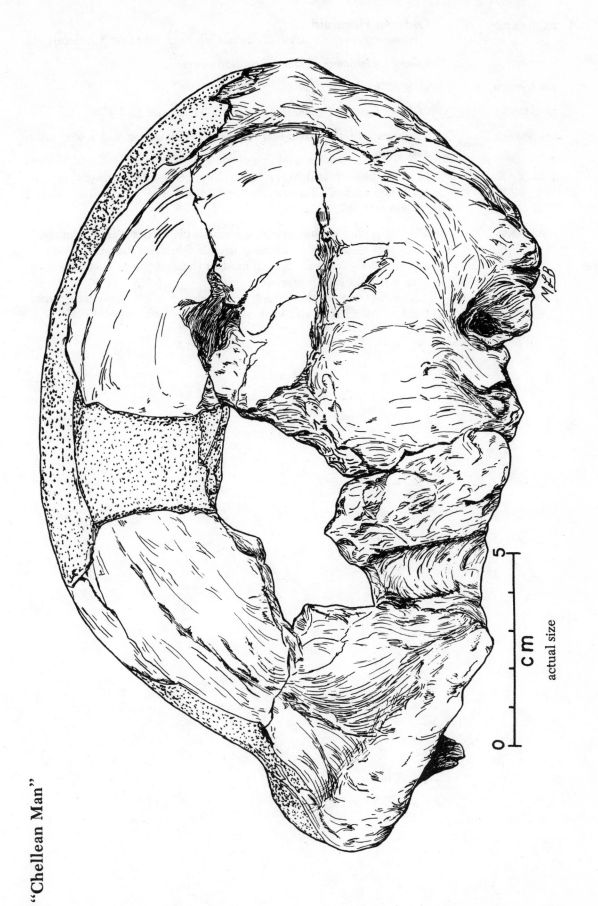

"Chellean Man"

5

cm

actual size

0

69

DESIGNATION:	**Olduvai Hominid 13** (*Homo erectus*, occasionally called *"Homo habilis"* or "Cinderella").
DATING:	Lower Pleistocene, about 1.5 million years ago.
DISCOVERED BY:	L. S. B. Leakey (1963).
LOCALE:	Olduvai Gorge, Tanzania (Site MNK II, Upper Bed II).
MATERIALS:	Fragments of the back end of a skull, the upper and lower jaws and teeth.
SOURCE:	L. S. B. Leakey and Mary D. Leakey, 1964, recent discoveries of fossil hominids in Tanganyika: at Olduvai and near Lake Natron. *Nature* 202:5–7.
COMMENT:	This is another of the various specimens claimed as *"Homo habilis."* Since it is measurably indistinguishable from the Pithecanthropines and relatively less robust than most of them, it seems best to regard it as a female *Homo erectus*. Drawn from a photograph and printed with permission from Dr. L. S. B. Leakey and Prof. P. V. Tobias.

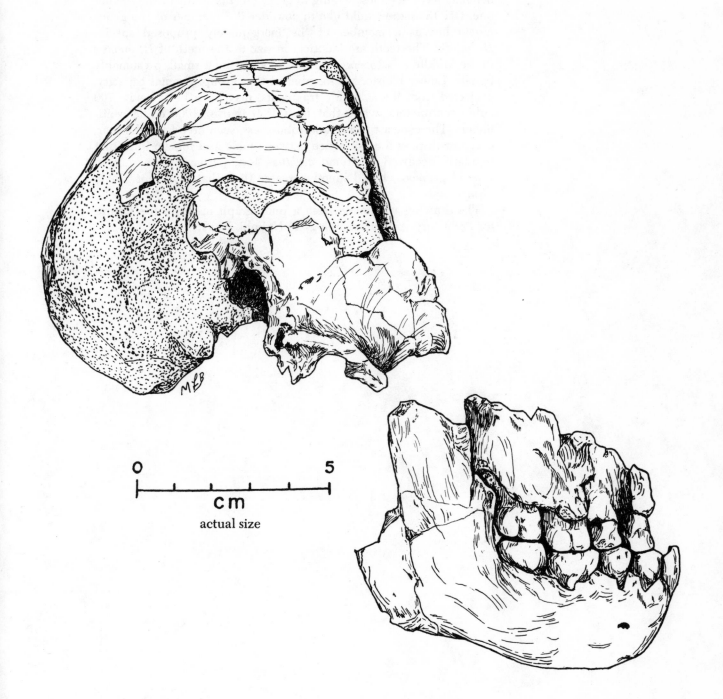

0   5

cm

actual size

DESIGNATION:

COMMENT:

Erectus Sexual Dimorphism

These skulls are from the middle and upper parts of Bed II, Olduvai Gorge, 1.3 to 1.5 million years ago. No one doubts that the larger one, Olduvai Hominid 9, is a representative of *Homo erectus*. Scholars have been less willing to grant *erectus* status to the smaller one, OH 13. Some would like to consider it a member of the genus *Homo* but as a member of the inadequately proposed species *H. habilis*. The teeth are identical in size to the teeth of *H. erectus* in the Middle Pleistocene, and this may just be a small, presumably female, Lower Pleistocene member of that group. Cranial capacity, projected from the available fragments, may have been below 700 cubic centimeters, while that for OH 9 is over 1000 cubic centimeters. The contrast in size and robustness, seen here from the rear, may just display the difference between male and female form that regularly occurred in *Homo erectus*—less, perhaps, than that in *Australopithecus*, but distinctly greater than that in recent *Homo sapiens*.

The drawing was made from a photograph made in the Centre for Prehistory and Palaeontology in Nairobi, Kenya.

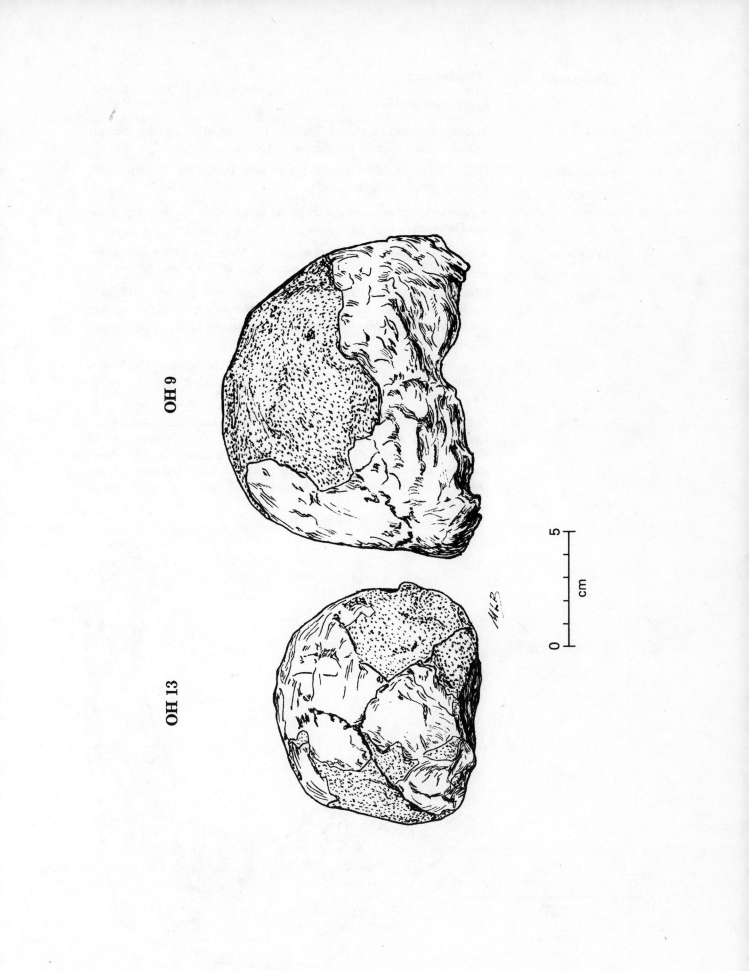

OH 9

OH 13

cm

5

0

73

DESIGNATION:	**"Heidelberg"** (*Homo erectus*, the Mauer jaw, originally called *"Homo heidelbergensis"*).
DATING:	Middle Pleistocene, most likely an interstadial in the second (Mindel) glaciation in the Alpine sequence.
DISCOVERED BY:	Gravel pit workers who turned it over to Prof. Otto Schoetensack, German paleontologist (1907).
LOCALE:	A commercial sand and gravel pit at Mauer near Heidelberg, western Germany.
MATERIAL:	A single complete *H. erectus* mandible. More recently crude stone tools have been found from the same level of the gravel pit.
SOURCE:	Otto Schoetensack, 1908, *Der Unterkiefer des Homo Heidelbergensis aus dem Sanden von Mauer bei Heidelberg. Ein Beitrag zur Paläontologie des Menschen.* Leipzig: Wilhelm Engelmann Verlag, 67 pp.
	Alfred Rust, 1956, *Artefakte aus der Zeit des Homo Heidelbergensis in süd-und Norddeutschland.* Bonn: Rudolf Habelt Verlag, 43 pp.
COMMENT:	Initially called *"Homo heidelbergensis,"* it is now regarded as belonging to a northwestern representative of *Homo erectus*. As is true for other *erectus* fossils, the teeth are not distinguishable from the later Neanderthals, although the jaw itself is more robust. Therefore its identification as *Homo erectus* is largely the consequence of its dating. The drawings were made from a cast.

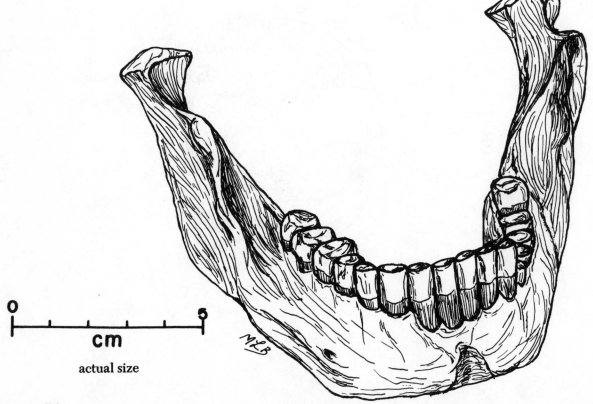

0   5

cm

actual size

74

DESIGNATION: **"Heidelberg."**

COMMENT: The side view shows the very broad ascending ramus and the generally robust and chinless character of the mandible.

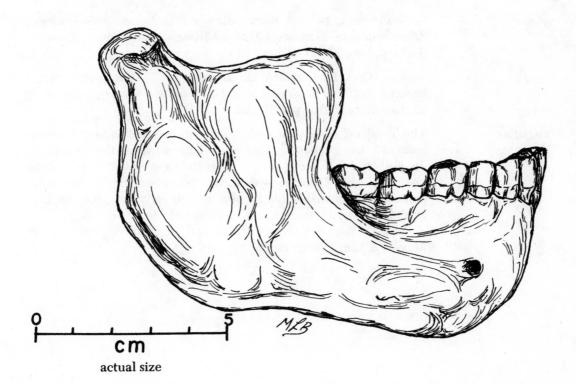

0   5

cm

actual size

DESIGNATION:	**"Ternifine"** (*Homo erectus,* originally referred to as "Atlanthropus mauritanicus.")
DATING:	Middle Pleistocene.
DISCOVERED BY:	C. Arambourg, French anthropologist (1954 and 1955).
LOCALE:	Ternifine, just southeast of Mascara near the coast of Algeria.
MATERIALS:	Three mandibles, a single skull bone (parietal), and some teeth, along with crude stone tools.
SOURCES:	C. Arambourg, 1955, A recent discovery in human paleontology: *Atlanthropus* of Ternifine (Algeria). *American Journal of Physical Anthropology* 13:191–202. ————, 1963, Le Gisement de Ternifine. I. L'Atlanthropus mauritanicus. *Archives de l'Institut de Paléontologie Humaine.* Mémoire 32, Deuxième Partie. pp. 37–190.
COMMENT:	The Ternifine specimens were initially called "*Atlanthropus mauritanicus.*" Unfortunately, this was done without the benefit of detailed study and comparison. As is the case with the Heidelberg mandible, the Ternifine specimens themselves could be either *erectus* or Neanderthal, but in view of their early date, they are best regarded as northwest African members of *Homo erectus.* Drawn after Arambourg 1963 and reproduced courtesy of Masson & Cie, Paris, France.

"Ternifine"

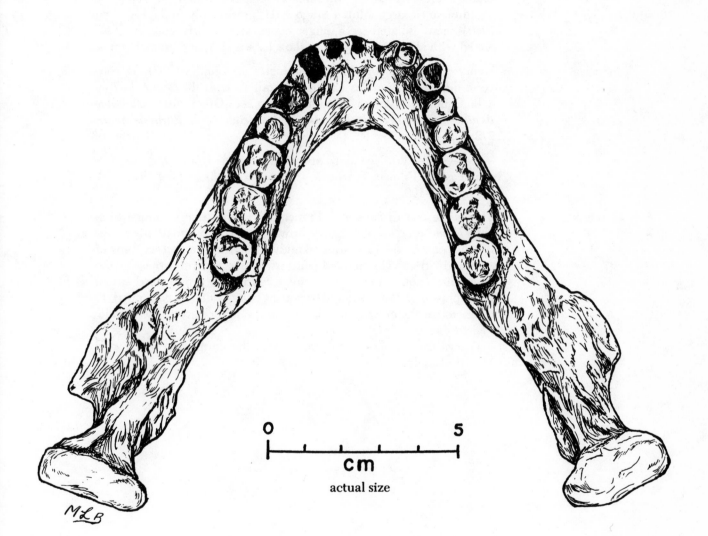

0   5

cm

actual size

MℒB

77

DESIGNATION:	**"Arago XXI"** (*Homo erectus*, "Tautavel Man").
DATING:	Middle Pleistocene, 200,000 years ago.
DISCOVERED BY:	Henry and Marie-Antoinette de Lumley, French prehistorians (1971).
LOCALE:	A cave, La Caune de l'Arago, just above the village of Tautavel, near the eastern end of the French Pyrenees.
MATERIALS:	A facial skeleton and frontal bone, found in 1971, a mandible of a small adult (Arago II, presumably female, found in 1969), a mandible of a large adult (Arago XIII, presumably male, found in 1970), many isolated teeth, finger bones, and skull fragments. Associated with abundant Acheulian tools (see p. 58), and animal bones.
SOURCES:	Henry de Lumley and Marie-Antoinette de Lumley, 1971, Découverte des restes humains anténéandertalien datés du début du Riss à la Caune de l'Arago (Tautavel, Pyrenées-Orientales). *Comptes Rendus des Séances de l'Académie des Sciences de Paris*, Série D, 272:1739–1742.
	———, 1974, Pre-Neanderthal human remains from Arago cave in southeastern France. *Yearbook of Physical Anthropology* 1973, Vol. 17, pp. 162–168.
COMMENT:	This is the first discovery in France of material that is unquestionably *Homo erectus*. The heavy brow ridge and prognathous tooth-bearing part of the face are strikingly similar to the *erectus* finds of Asia and Africa. The marked contrast in size and robustness of the mandibles (not pictured here) suggests that the pronounced sexual dimorphism of the earlier African *erectus* evidence continued to characterize the *erectus* populations in the latter part of the Middle Pleistocene. Drawn from a photograph taken with the permission of Dr. and Mrs. de Lumley.

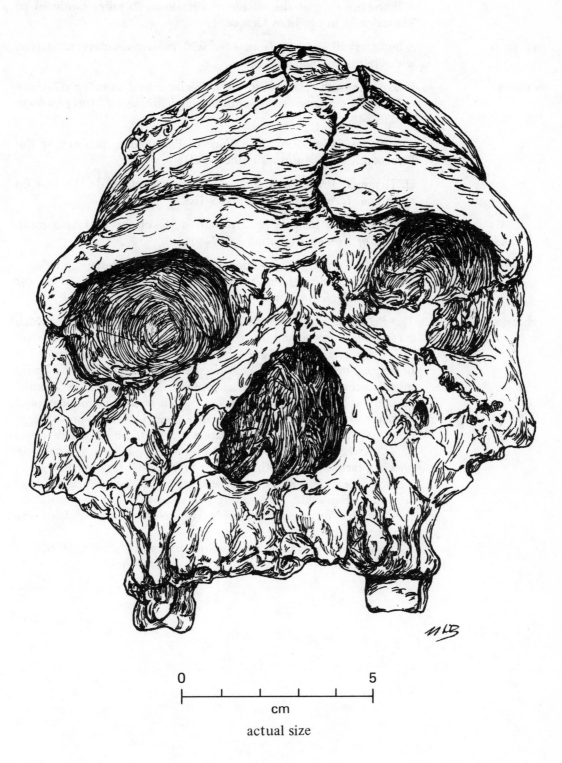

0   5

cm

actual size

DESIGNATION:	**"Petralona"** (*Homo erectus* ?)
DATING:	Middle Pleistocene?
DISCOVERED BY:	A group of Greek cave explorers (1959).
LOCALE:	A limestone cave at the village of Petralona, 20 miles southeast of Thessaloniki in northern Greece.
MATERIALS:	A human skull and face (encrusted with calcareous material) mixed with Pleistocene mammal bones.
SOURCES:	P. Kokkoros and A. Kanellis, 1960, Découverte d'un crâne d'homme paléolithique dans la péninsule chalcidique. *L'Anthropologie* 64:438–446. Aris N. Poulianos, 1967, The place of Petralonian man among the Palaeoanthropoi. *Anthropos* 19:216–221. H. Hemmer, 1972, Notes sur la position phylétique de l'homme de Petralona. *L'Anthropologie* 76:155–162. C. B. Stringer, 1974, A multivariate study of the Petralona skull. *Journal of Human Evolution* 3:397–404. Rupert I. Murrill, 1975, A comparison of the Rhodesian and Petralona upper jaws in relation to other Pleistocene hominids. *Zeitschrift für Morphologie und Anthropologie* 66:176–187.
COMMENT:	The Petralona skull, like the Rhodesian skull (see page 87), which it resembles to an extent, is evidently earlier than the Upper Pleistocene date originally assigned to it. In fact, it may date from the beginning of the Middle Pleistocene. Morphologically one could justify either an *erectus* or a *sapiens* designation. The cranial capacity is 1220 cubic centimeters, which is low for *sapiens* and high for *erectus* but could occur in either. The huge brow ridge, thick cranial bones, and large palate ally it with the known members of *Homo erectus* in the Middle Pleistocene. It can serve as an ancestor for the Neanderthal form of *sapiens* that was to come along in the Upper Pleistocene, and the doubts that exist among professional anthropologists concerning just what to call it only point up the fact that *erectus* ultimately gave rise to *sapiens* without any conspicuous break. Drawn from a photograph in Kokkoros and Kanellis, 1960, and reproduced courtesy of Masson & Cie., Paris, France.

"Petralona"

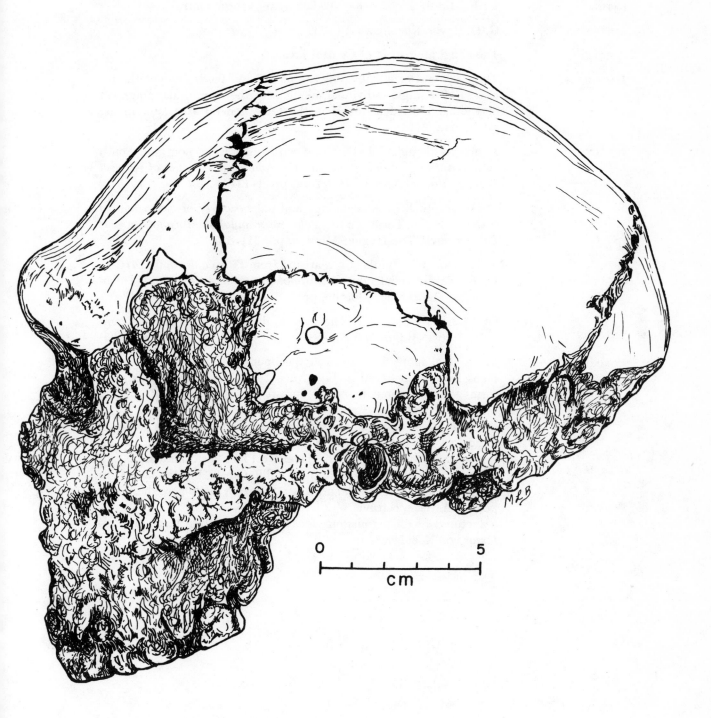

0   5
cm

DESIGNATION:	**"Pithecanthropus IV"** (with "Pithecanthropus mandible B," *Homo erectus,* variously referred to as *"Pithecanthropus robustus," "Pithecanthropus modjokertensis," "Pithecanthropus erectus* IV," and "Sangiran").
DATING:	Early Middle Pleistocene, 700,000 years ago or more.
DISCOVERED BY:	G. H. R. von Koenigswald (1939).
LOCALE:	The Sangiran district of Central Java.
MATERIALS:	The back end of a skull plus the palate and teeth (except the incisors) of a single individual, along with the mandibular fragment of another individual found in the same general area later in the same year.
SOURCES:	Franz Weidenreich, 1945, Giant early man from Java and South China. *Anthropological Papers of the American Museum of Natural History,* Vol. 40, Part I. New York. pp. 1–134.
	Teuku Jacob, 1975, Morphology and paleoecology of early man in Java. In R. H. Tuttle (ed.), *Paleoanthropology: Morphology and Paleoecology.* The Hague: Mouton, pp. 311–325.
	S. Sartono, 1975, Implications arising from *Pithecanthropus VIII.* In R. E. Tuttle (ed.), *Paleoanthropology: Morphology and Paleoecology.* The Hague: Mouton, pp. 327–360.
	Gert-Jan Bartstra, 1978, The age of the Djetis beds in east and central Java. *Antiquity* 52:56–58.
COMMENT:	Remains of more than a dozen specimens of *Homo erectus* specimens have been found in Java since 1960. Their appearance conforms to expectations based on the previous Javanese finds and on material of the same stage in Africa, Europe, and China. Few of the finds have been made *in situ* by trained scientists and problems of geology and dating remain to be solved. Potassium-argon determinations have been run for the Javanese Plio-Pleistocene strata and these vary between 500,000 and 1.9 million years, but the layers are more than 1,000 feet thick and it is not certain just where the fossils belong in that spectrum. Pictured (with permission) is a recent restoration made by Professor von Koenigswald.

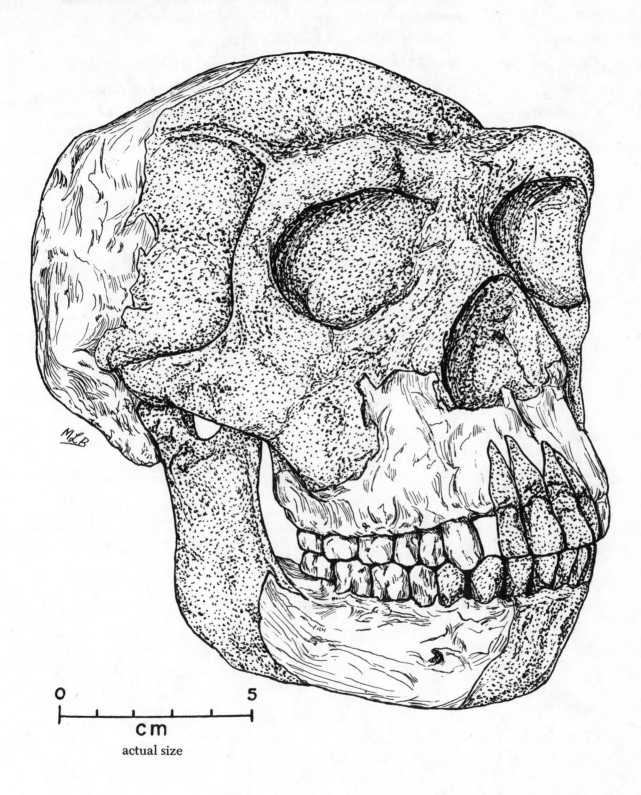

0   5

cm

actual size

DESIGNATION:	**"Saldanha"** (*Homo erectus*, "Hopefield," once erroneously called *"Homo saldanhanensis"*)
DATING:	Middle Pleistocene, perhaps 500,000 years ago.
DISCOVERED BY:	Ronald Singer and Keith Jolly, South African anthropologists (1953).
LOCALE:	At Elandsfontein near Hopefield, Saldanha Bay, South Africa.
MATERIALS:	A skull cap and a mandible found too far apart to have belonged to the same individual, along with Acheulian stone tools (see page 58) and the bones of large-sized Pleistocene mammals very like those found at Olduvai Gorge between Beds II and IV.
SOURCES:	Ronald Singer, 1954, The Saldanha skull from Hopefield, South Africa. *American Journal of Physical Anthropology* 12:345–362. Richard G. Klein, 1977, The ecology of early man in southern Africa. *Science* 197:115–126.
COMMENT:	In form, the Saldanha skull cap is so like the "Rhodesian" skull (see page 87) that it was long thought to be late Pleistocene in spite of the geological evidence. Recent work has shown that both are probably Middle Pleistocene African representatives of *Homo erectus*. Drawn from a photograph and printed with the permission of E. M. Shaw and Q. B. Hendey, South African Museum, Cape Town.

84

"Saldanha"

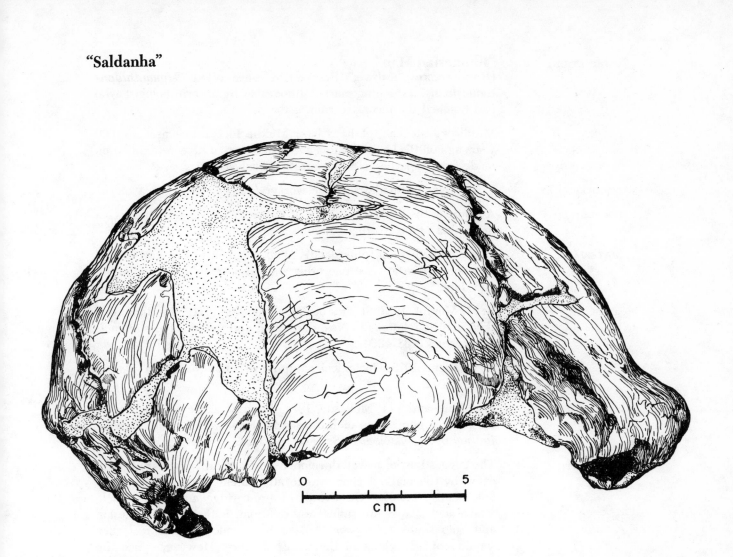

DESIGNATION:	**"Rhodesian Man"** (*Homo erectus,* "Kabwe," "Broken Hill," once called *"Cyphanthropus rhodesiensis"* (stooping man of Rhodesia) by an ornithologist who had botched the reconstruction of the pelvis.)
DATING:	Middle Pleistocene, perhaps later Middle Pleistocene near 130,000 years ago on the basis of fauna and also on a tentative determination made from aspartic acid racemization analysis.
DISCOVERED BY:	T. Zwiglaar, a Rhodesian mine worker (1921).
LOCALE:	A fissure in the Broken Hill Mine at Kabwe, 75 miles north of Lusaka in Zambia (then Northern Rhodesia).
MATERIALS:	One complete skull, plus palate, skull fragments, and post-cranial fragments from at least one other individual of unknown affinities and date.
SOURCES:	G. M. Morant, 1928, Studies of Paleolithic Man, III: The Rhodesian skull and its relations to Neanderthaloid and modern types. *Annals of Eugenics* 3:337–360.
	R. G. Klein, 1973, Geological antiquity of Rhodesian Man. *Nature* 244:311–312.
	J. L. Price and T. I. Molleson, 1974, A radiographic examination of the left temporal bone of Kabwe man, Broken Hill mine, Zambia. *Journal of Archaeological Science* 1:285–289.
COMMENT:	The suggestions of a well-developed post-cranial musculature indicated by the marked neck muscle attachments remind one of the Solo skulls. Some of the details of brow ridge and neck muscle attachment shapes are sufficiently different, both in the Rhodesian and Solo skulls (see page 99), so that some scholars feel they should not all belong in the same category. However, since the general evolutionary significance is the same despite the individual idiosyncrasies, it makes better sense to consider them respectively as African and Asian representatives of the transition from *Homo erectus* to a Neanderthal-like form of *Homo sapiens.* In capacity, the Kabwe skull is just under 1300 cubic centimeters which is a little small for *sapiens* and a little large for *erectus.* The huge face and brow ridges and the indications of powerful musculature are Middle Pleistocene characteristics.

Drawn from a photograph and reproduced with the permission of D. R. Brothwell, British Museum (Natural History).

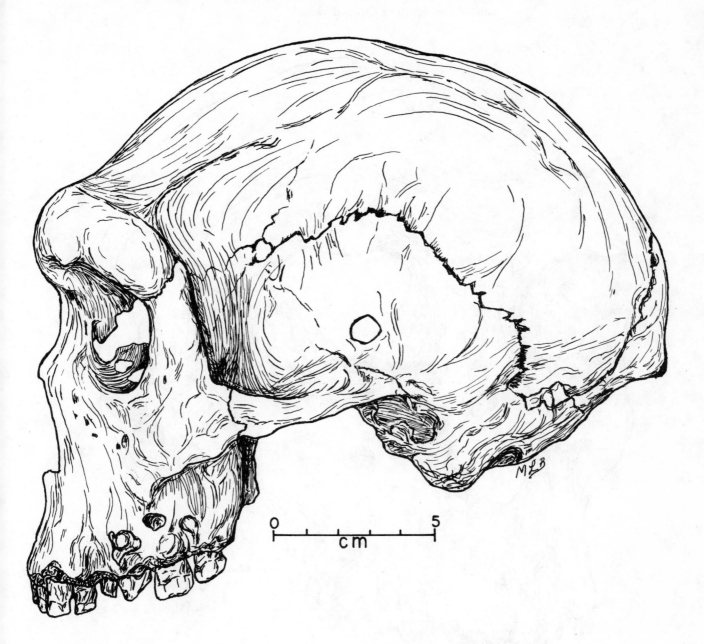

DESIGNATION: **"Rhodesian Man"**
(Homo erectus ? cf. *sapiens* ?)

COMMENT: This view graphically displays the enormous brow ridges of the Rhodesian skull. As an additional note, it is interesting to learn that the Rhodesian dentition shows the first known example of extensive tooth decay. In fact, the resultant secondary infection may have caused the death of this individual.

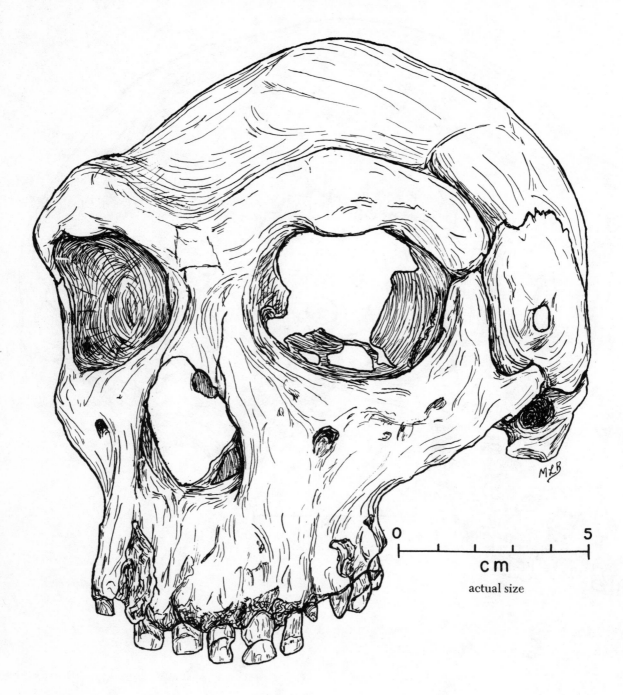

0   5

c m

actual size

EARLY HOMO SAPIENS: THE ERECTUS-SAPIENS TRANSITION

The principal significant change from *Homo erectus* is the increase in brain size, allowing us to classify the Neanderthals in the same species that includes modern man, *Homo sapiens*. The retention of the large Middle Pleistocene jaws and teeth leads some to classify them as subspecifically distinct from modern man, properly being called *Homo sapiens neanderthalensis*. This is the classification followed here.

When cultural remains are found in association with the skeletal material, they are usually of the Mousterian variety or its Middle Palaeolithic equivalent in parts of the world other than Europe or the Middle East.

DESIGNATION:	**"Steinheim"**
	(*Homo sapiens* ?)
DATING:	Middle Pleistocene, 200,000 (?) years ago.
DISCOVERED BY:	F. Berckhemer, German paleontologist (1933).
LOCALE:	In a gravel pit at Steinheim, 20 miles north of Stuttgart, West Germany.
MATERIALS:	One somewhat distorted skull, including much of the face but missing the forward part of the dental arch and palate.
SOURCES:	H. Weinert, 1936, Der Urmenschenschädel von Steinheim. *Zeitschrift für Morphologie und Anthropologie* 35:413–518.
	F. Clark Howell, 1960, European and northwest African Middle Pleistocene hominids. *Current Anthropology* 1:195–232.
COMMENT:	Once called *"Homo steinheimensis,"* it is now evident that this should not be given a species all to itself. However, it is not clear in just what species it should be included. The cranial capacity, about 1100 cubic centimeters, is large for *Homo erectus* and small for *Homo sapiens*. Muscle markings are not pronounced but the brow ridge is imposing. The third molars are noticeably reduced but the palate apparently was large, particularly toward the front end which is unfortunately missing. Altogether, this appears to represent the transition from *Homo erectus* to the Neanderthal form of *Homo sapiens*, with some of the "modern" features being due to the fact that it was probably a female.

The drawing is of a cast.

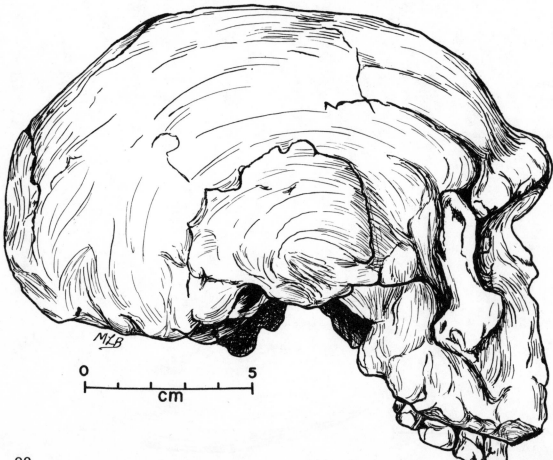

0   5
cm

"Steinheim"

We include this view to show the extent of the warping which the skull had undergone after death. Some of the previous claims for the "modern" appearance of Steinheim may not have taken this into account. Note the breakage at the forward end of the palate. If the face had been preserved unbroken, it is apparent that the jaws and teeth would have projected in a decidedly nonmodern manner.

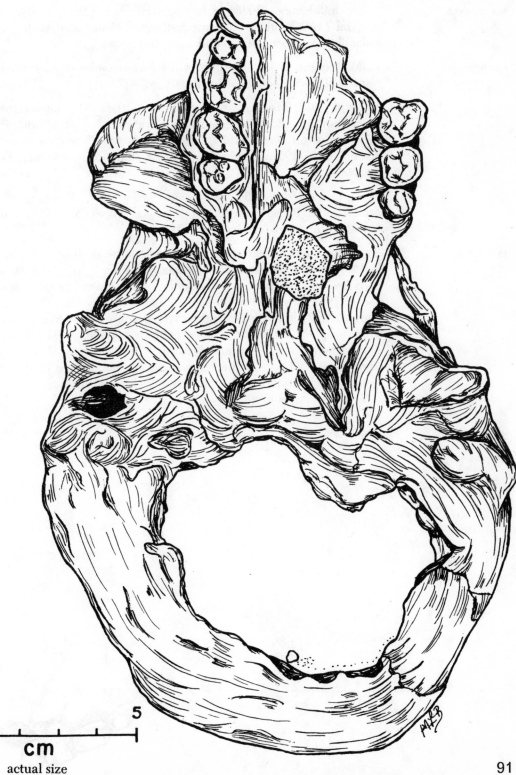

0   5

cm

actual size

DESIGNATION:	**"Swanscombe"** (*Homo sapiens* ?)
DATING:	Middle Pleistocene, 150,000 to 200,000 (?) years ago.
DISCOVERED BY:	A. T. Marston, English dentist and amateur archeologist (1935 and 1936) and J. Wymer, English archeologist (1955).
LOCALE:	A gravel pit on the banks of the Thames River below London, England.
MATERIALS:	An occipital bone (1935), a left parietal bone (1936), and a right parietal (1955), all of the same individual. Associated with Lower Palaeolithic (Acheulian) hand axes and flakes.
SOURCES:	G. M. Morant, 1938, The form of the Swanscombe skull. *Journal of the Royal Anthropological Institute* 68:67–97.
	S. Sergi, 1953, Morphological position of the "Prophaneranthropi" (Swanscombe and Fontéchevade). Translated and reprinted in W. W. Howells (ed.), 1962, *Ideas on Human Evolution: Selected Essays, 1949–1961*. Cambridge, Mass.: Harvard University Press, pp. 507–520.
	C. D. Ovey (ed.), 1964, *The Swanscombe Skull: A Survey of Research on a Pleistocene Site*. Occasional Paper No. 20, London: Royal Anthropological Institute, 215 pp.
COMMENT:	Arguments about the form of the Swanscombe skull have continued right from the beginning, with many attempts to claim it as a "modern" form. However, the only comparative metrical studies (Sergi 1953, and Weiner and Campbell in Ovey ed. 1964) clearly show that its affinities are properly with the Neanderthal form of *Homo sapiens* rather than more modern populations. Lacking the frontal bone, face, jaws, and teeth, it is hard to say. Although disagreement continues, it may represent another female of the same general population that included Steinheim. The drawing is of a cast.

"Swanscombe"

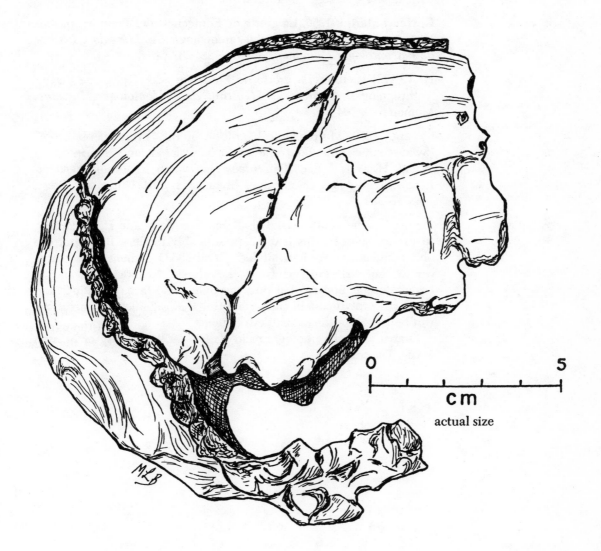

0   5

cm

actual size

DESIGNATION:	**"Fontéchevade"** (*Homo sapiens* ?)
DATING:	Middle Pleistocene, 115,000 (?) years ago.
DISCOVERED BY:	Mlle. G. Henri-Martin, French prehistorian (1947).
LOCALE:	The collapsed cave of Fontéchevade, Charente, southwestern France.
MATERIALS:	A small fragment of frontal bone from one individual (perhaps immature) and an incomplete skull cap from an adult. Associated with Lower Palaeolithic cultural remains.
SOURCES:	G. Henri-Martin, 1956, La grotte de Fontéchevade: Prémière partie: Historique, fouilles, stratigraphie, archéologie. *Archives de L'Institut de Paléontologie Humaine*, Mémoire 28, 287 pp.
	H. V. Vallois, 1959, La grotte de Fontéchevade: Deuxième partie: Anthropologie. *Archives de L'Institut de Paléontologie Humaine*, Mémoire 29, 164 pp.
	S. Sergi, 1953, Morphological position of the "Prophaneranthropi" (Swanscombe and Fontéchevade). Translated and reprinted in W. W. Howells (ed.), 1962, *Ideas on Human Evolution: Selected Essays, 1949–1961*. Cambridge, Mass.: Harvard University Press, pp. 507–520.
COMMENT:	Pictured is the controversial skull cap Fontéchevade II. According to French sources, this is of unquestionably modern form, but the only comparative study available (Sergi 1953) demonstrates that it cannot be distinguished from the classic Neanderthals and is definitely not modern. Lacking the diagnostic face, dentition, and basal parts of the skull, it is doubtful that significant conclusions can be based on the available fragments. Drawn after the photograph in Vallois 1959, courtesy of Masson & Cie, Paris, France.

94

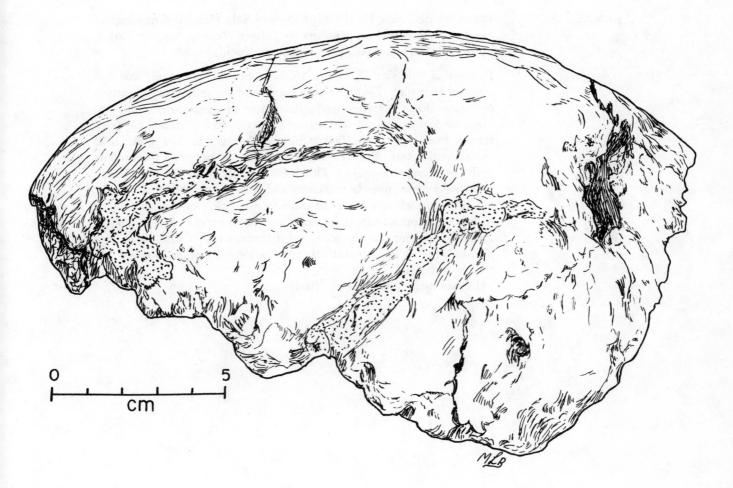

0   5

cm

DESIGNATION:	**"Solo Man"** (*Homo sapiens neanderthalensis* ? Sometimes called *"Homo solo-ensis," "Javanthropus,"* or Ngandong man.)
DATING:	Late Middle Pleistocene ?
DISCOVERED BY:	C. ter Haar, Dutch geologist, and W. F. F. Oppenoorth, Dutch engineer (1931–1933).
LOCALE:	Ngandong on the banks of the Solo River in central Java.
MATERIALS:	Eleven skulls minus faces, jaws, and teeth. Also found were two tibiae (shin bones).
SOURCE:	Franz Weidenreich, 1951, Morphology of Solo Man. *Anthropological Papers of the American Museum of Natural History*, Vol. 43, Part 3, pp. 1–290. Introduction by G. H. R. von Koenigswald.
COMMENT:	Pictured is Solo skull XI. Until 1958, when the Mapa skull was discovered (see page 134), the Solo finds were the only representatives of human form between unquestioned *erectus* specimens and obvious modern humans anywhere east of Asia Minor. Cranial capacity ranges from just over 1000 cubic centimeters to just over 1250 cubic centimeters, with the average falling in between the *erectus* and Neanderthal averages. This measurement, plus the very heavy brow ridges and muscle markings and other morphological details, led Weidenreich and others to consider the Solo skulls as belonging to a population intermediate between *Homo erectus* and Neanderthals. The dating is not precise, but it seems a reasonable interpretation to offer, at least tentatively. Archeological associations are not well worked out. Drawn after Weidenreich, 1951.

96

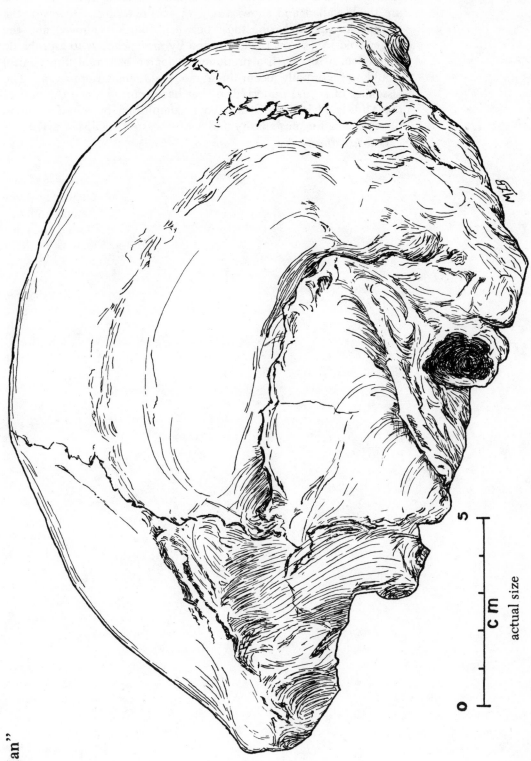

"Solo Man"

cm

actual size

5

0

97

DESIGNATION: **Late Pithecanthropine/Early Neanderthal Sexual Dimorphism**

COMMENT: Since the material from Ngandong consisted mainly of skulls, the identification of sex cannot be absolutely certain. All of the skulls are thick walled and show strong muscle markings. However, the contrast in size and robustness between Solo VI (above) and the larger Solo IX (below) is attributed by most scholars to a probable difference in sex. Such a pronounced degree of sexual dimorphism would not normally occur in a modern human population. The lessening of sexual dimorphism that has occurred since the end of the Middle Pleistocene has been accomplished by a reduction in the evidence for muscularity and robustness especially in males.

Drawn from casts.

"Solo VI"

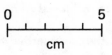

"Solo IX"

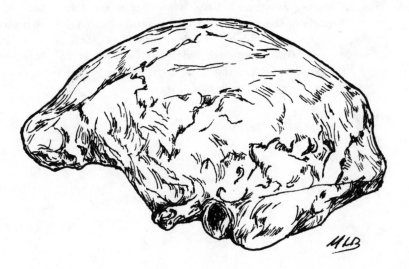

99

EARLY HOMO SAPIENS: NEANDERTHALS

Neanderthal status was achieved by the time brain size had reached fully modern levels. This occurred before the Middle Pleistocene faces and teeth began to reduce toward their modern levels and before the evidence for muscularity and skeletal robustness began the reduction that ultimately produced modern form. At the beginning of the Neanderthal Stage, braincase reinforcements and brow ridges are still of *erectus* size, and there is some reason to claim that the incisors and the front end of the dental arch reach the maximum size achieved during the whole course of hominid evolution.

In general, the Neanderthal Stage was reached in the last interglacial more than 100,000 years ago. Somewhere between 35,000 and 40,000 years ago the reductions in muscularity, skeletal robustness, and face size that had been gradually accumulating since the onset of the last glaciation 80,000 years ago led to the appearance of what we regard as modern human form. The change was very gradual and it did not occur at the same rate of speed in all parts of the world. Just which forms qualify for "primitive" modern status and which are relegated to modified Neanderthal status is an arbitrary matter of judgment, but it engages professional anthropologists in seemingly endless arguments which, in perspective, are really not very important.

NEANDERTHAL SITES

(Where the most complete and/or important specimens have been found)

1. Neanderthal
2. Spy
3. Ehringsdorf
4. La Chapelle-aux-Saints
5. Le Moustier
6. La Ferrassie
7. La Quina
8. Gibraltar
9. Saccopastore
10. Monte Circeo
11. Krapina
12. Teshik Tash (Uzbek S. S. R.)
13. Shanidar (Iraq)
14. Mount Carmel
15. Haua Fteah (Libya)
16. Jebel Irhoud (Morocco)
17. Diré Dawa (Ethiopia)
18. Cave of the Hearths (South Africa)
19. Mapa (China)

▓ Distribution of cold-adapted Mousterian cultural elements in the early Upper Pleistocene.

▨ Distribution of contemporary but less specialized "Mousterioid" or Middle Palaeolithic cultural remains.

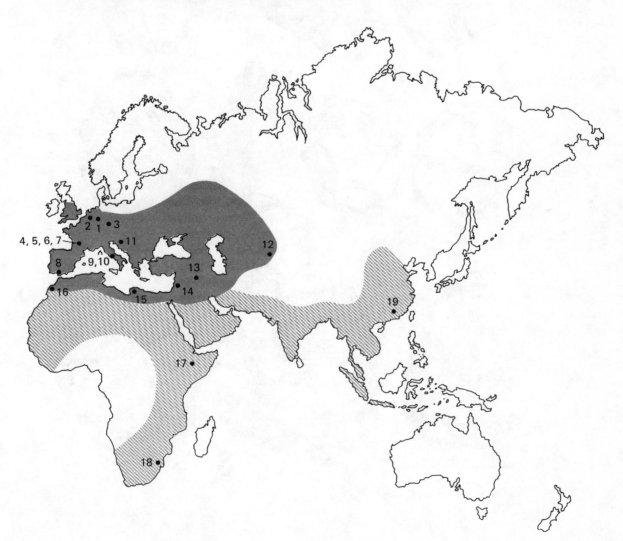

MOUSTERIAN TOOLS: *a, b*) Side scrapers from Le Moustier, southwestern France; *c*) Disc-core from Le Moustier; *d*) Point from Le Moustier; *e*) Anvil or hammerstone from Gibraltar; *f*) Biface (hand axe) from Le Moustier; *g*) Biface (from Kent's Cavern, Torquay, England; *h*) Oval flake tool from Kent's Cavern.

(Reprinted from Oakley 1950, by permission of the Trustees of the British Museum [Natural History].)

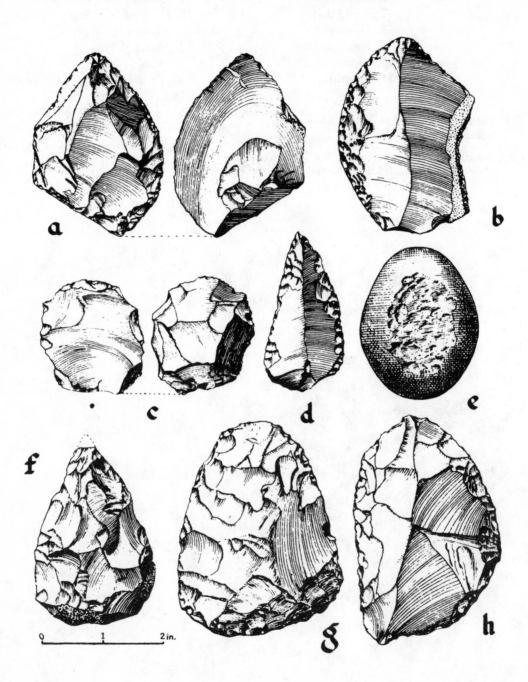

DESIGNATION:	**"Neanderthal"**
	(*Homo sapiens neanderthalensis,* also referred to as *"Homo neanderthalensis," "Homo primigenius," "Homo mousteriensis,"* even *"Palaeanthropus"* and other names.)
DATING:	Upper Pleistocene, but because of the accidental and informal means of discovery, no precise date can be assigned.
DISCOVERED BY:	Quarry workers, later recognized by J. K. Fuhlrott, German high school science teacher (1856).
LOCALE:	A limestone cave in the Neander valley near Düsseldorf, western Germany.
MATERIALS:	Originally probably a complete skeleton of which only the major long bones, ribs and skull cap were preserved.
SOURCES:	Gustav Schwalbe, 1901, Der Neanderthalschädel. *Bonner Jahrbücher,* No. 106, pp. 1–72.
	G. H. R. von Koenigswald (ed.), 1958, *Hundert Jahre Neanderthaler: Neanderthal Centenary, 1856–1956.* Utrecht, Netherlands: Kemink en Zoon N. V., 325 pp.
COMMENT:	The specimen on pages 74 and 75, the original "Neanderthal," was the first non-modern human fossil recognized as such. It has given its name to an entire stage in human evolution.
	The drawings were made from a cast.

"Neanderthal"

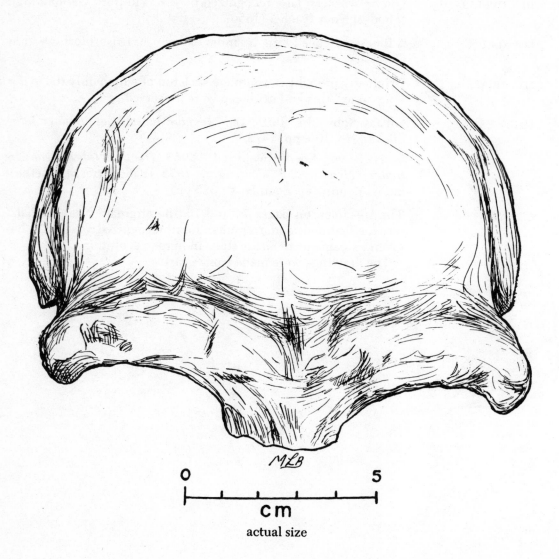

cm

actual size

104

DESIGNATION: **"Neanderthal"**

COMMENT: This is a side view of the same skull. The heavy brow ridge and sloping forehead are characteristic of male Neanderthals, although there is a good deal of variation from individual to individual. These features are less clear-cut in females.

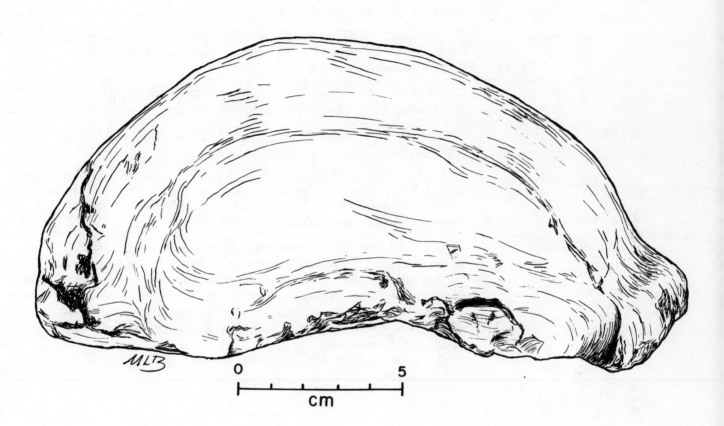

0   5

cm

DESIGNATION:	**"Ehringsdorf"**
	(*Homo sapiens neanderthalensis.*)
DATING:	Late in the last interglacial.
DISCOVERED BY:	Workers in Kämpfe's and Fischer's Quarries from 1908 to 1925, recognized by E. Lindig, preparator at the Weimar City Museum.
LOCALE:	Ehringsdorf, near Weimar, East Germany.
MATERIALS:	An adult jaw (Kämpfe's Quarry, 1914); a jaw, some teeth, and some postcranial fragments of a 10-year-old child (Kämpfe's Quarry, 1916); some other isolated fragments; and, most important, a reconstructible skull in Fischer's Quarry (1925).
SOURCES:	H. Virchow, 1920, *Die menschlichen Skelettreste aus dem Kämpfe'-schen Bruch im Travertin von Ehringsdorf bei Weimar.* Jena: Gustav Fischer Verlag, 141 pp.
	Franz Weindenreich (ed.), 1928, *Der Schädelfund von Weimar-Ehringsdorf.* Jena: Gustav Fischer Verlag, 204 pp.
	Otto Kleinschmidt, 1931, *Der Urmensch.* Leipzig: Quelle and Meyer Verlag, 156 pp.
	G. Behm-Blancke, 1958, Umwelt, Kultur und Morphologie des eem-interglazialen-Menschen von Ehringsdorf bei Weimar. In G. H. R. von Koenigswald (ed.), *Hundert Jahre Neanderthaler.* Utrecht, Netherlands: Kemink en Zoon N. V., pp. 141–150.
COMMENT:	Claims have been made concerning the supposed "high vaulting" of the 1925 skull suggesting modern form. These are based on the reconstruction made by Weidenreich which evidently is in error. The reconstruction subsequently made by Kleinschmidt is more consistent and gives the skull full "classic" Neanderthal form: "bun-shaped" rear (occiput), heavy brow ridge, low vault, and relatively small mastoid processes.
	The drawing is of Kleinschmidt's reconstruction, after Behm-Blancke, 1958.

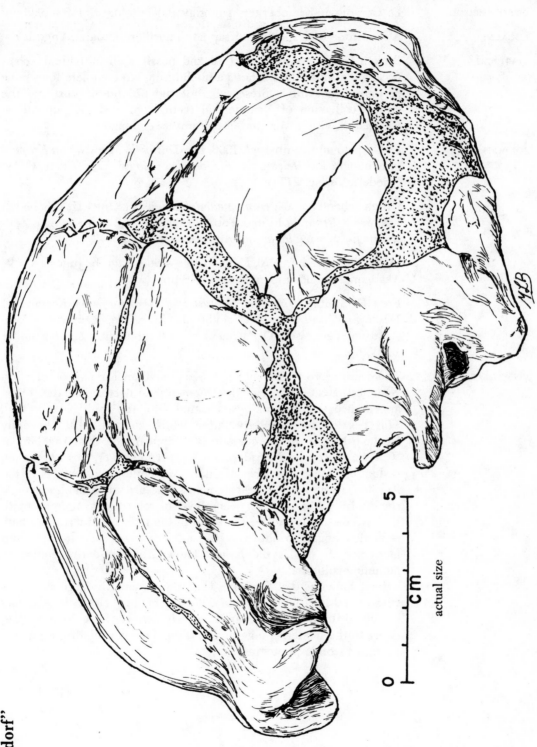

"Ehringsdorf"

5

cm

actual size

0

107

DESIGNATION:	**"Krapina"** (*Homo sapiens neanderthalensis.*)
DATING:	Late in the last interglacial, approximately 100,000 years ago.
DISCOVERED BY:	D. Gorjanović-Kramberger, Yugoslav paleontologist, 1899–1905.
LOCALE:	A sandstone rock shelter at Krapina in northern Croatia, Yugoslavia.
MATERIALS:	Broken fragments of at least 14 and possibly 15 individuals, comprising 9 adults and 5 infants and children. No complete long bone or unbroken skull is preserved, but the 270 teeth constitute the largest collection of Neanderthal teeth from a single population. Associated with a Mousterian cultural assemblage.
SOURCES:	D. Gorjanović-Kramberger, 1906, *Der Diluviale Mensch von Krapina in Kroatien: Ein Beitrag zur Paläoanthropologie.* Wiesbaden: C. W. Kriedel's Verlag, 277 pp.
	Ulrich Schaefer, 1964, *Homo neanderthalensis* (King) II. E-Schädel-Fragment, Frontale Fl und Torus-Fragment 37.2 von Krapina. *Zeitschrift für Morphologie und Anthropologie* 54:260–271.
	Mirko Malez (ed.), 1970, *Krapina 1899–1969.* Zagreb: Jugoslavenske Akademíje Znanosti i Umjetnosti, 216 pp.
	Fred H. Smith, 1976, *The Neanderthal Remains from Krapina: A Descriptive and Comparative Study.* Knoxville, Tennessee: University of Tennessee, Department of Anthropology, Report Number 15, 359 pp.
COMMENT:	There are more fragments from Krapina than for any other early Neanderthal site. Many of the fragments do not present the robust appearance that is usually associated with the term Neanderthal. This has led to an interpretation which suggests that the early Neanderthals were more modern in appearance than the late or "Classic" European Neanderthals. In fact, however, this impression is due to the large number of infants, juveniles, and females in the Krapina assemblage. The fragments that are identifiable as male show a fully Middle Pleistocene level of robustness. Average tooth size was distinctly greater than that of the later Neanderthals where reduction began to approach the modern range of variation, and incisor tooth size is the largest known for the entire course of hominid evolution. First shown is the "C skull." This is a moderately large skull (reconstructed length of 192 mm.), but the facial skeleton is relatively small and the brow ridges are relatively delicate for a Neanderthal, being in the modern range of variation. This is probably a female. Drawn from a photograph.

"Krapina" C skull

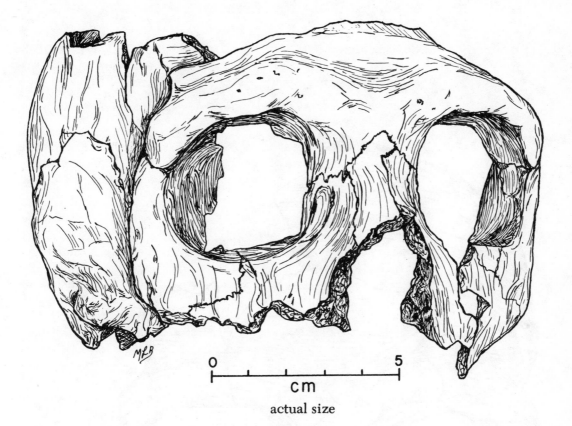

0   5

cm

actual size

DESIGNATION: **"Krapina" E skull**
(Homo sapiens neanderthalensis.)

COMMENT: This is the "E-Skull." The vault is relatively high and modern in shape but quite small (being 176 mm. in reconstructed length) with a good heavy Neanderthal brow ridge. This, too, is probably a female. Completing the skull outlines, it is clear that there is no foundation to early claims that the Krapina Neanderthal skulls were relatively broad in relation to their length.

The assemblage of pieces to produce the form pictured here was accomplished by the German anthropologist Ulrich Schaefer, working on the fragments in the collection at the National Geological and Paleontological Museum in Zagreb in 1957.

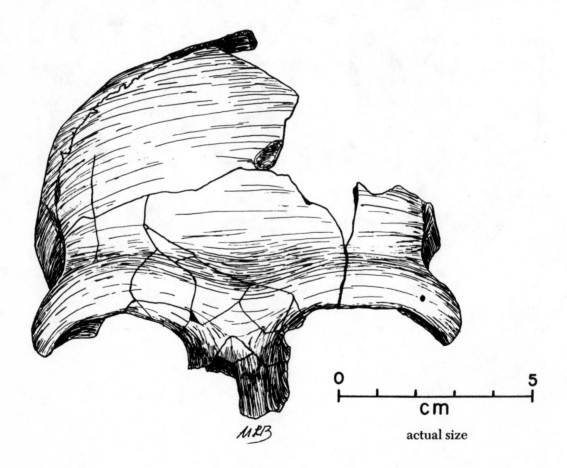

0   5
cm
actual size

DESIGNATION:	**"Krapina" D skull**
	(*Homo sapiens neanderthalensis.*)
COMMENT:	These fragments have been attributed to a single individual referred to as the "D-Skull." For the picture they were simply propped up with a piece of cloth. The shape therefore is very tentative, although the estimated length of over 200 mm. is reasonable. Certainly it was a large, thick skull with heavy brow ridges and strong neck muscle attachments, probably a male.
	Drawn from a photograph.

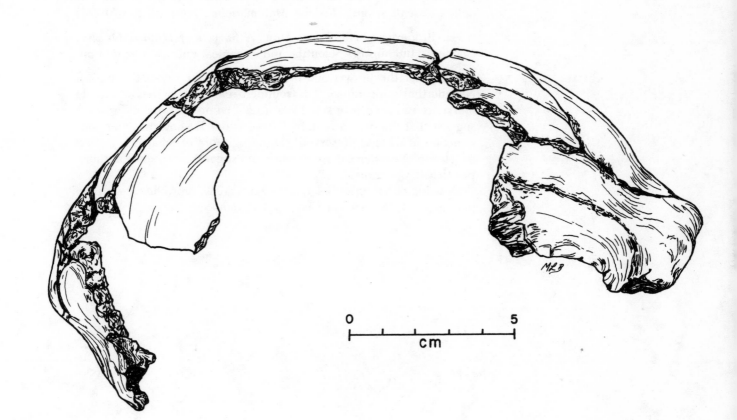

0   5

cm

DESIGNATION:	**"Saccopastore"** (*Homo sapiens neanderthalensis.*)
DATING:	Upper Pleistocene, end of the last interglacial.
DISCOVERED BY:	Gravel pit workmen (Saccopastore I, 1929); and H. Breuil, French archeologist, and A. C. Blanc, Italian archeologist, (Saccopastore II, 1935).
LOCALE:	A gravel pit on the outskirts of Rome, Italy.
MATERIALS:	One relatively complete skull damaged by the workman's pick: brow ridge broken off and a pick hole going clean through (Saccopastore I, 1929); and one broken skull consisting largely of face (palate, nose, and right eye socket) and part of the right side (Saccopastore II, 1935).
SOURCES:	S. Sergi, 1930, Il primo cranio del tipo di Neandertal scoperto in Italia nel suolo di Roma. *Bolletino Societa Geologica Italiana* 49: xxvii–xlv.
	S. Sergi, 1942, Sulla morfologia cerebrale del secondo paleantropo di Saccopastore. *Atti della Reale Accademia d'Italia. Rendiconti della Classe di Scienze, Fisiche, Matematiche e Naturali* 20:670–681.
	A. H. Broderick, 1948, *Early Man: A Survey of Human Origins.* London: Hutchinson's Scientific and Technical Publications, 288 pp.
COMMENT:	Breakage of the brow ridge of Saccopastore I reduces the typical Neanderthal appearance. The rounded rear, lack of heavy muscle markings and relatively small size (1200 cubic centimeters or less) suggest that this was a female. Saccopastore II shows heavier construction in what is preserved, exhibiting more of what is regarded as "classic" Neanderthal appearance and suggesting to some authorities that it was a male.
	Drawing of Saccopastore I, after Sergi, 1930, reproduced with the permission of the Societa Geologica Italiana.

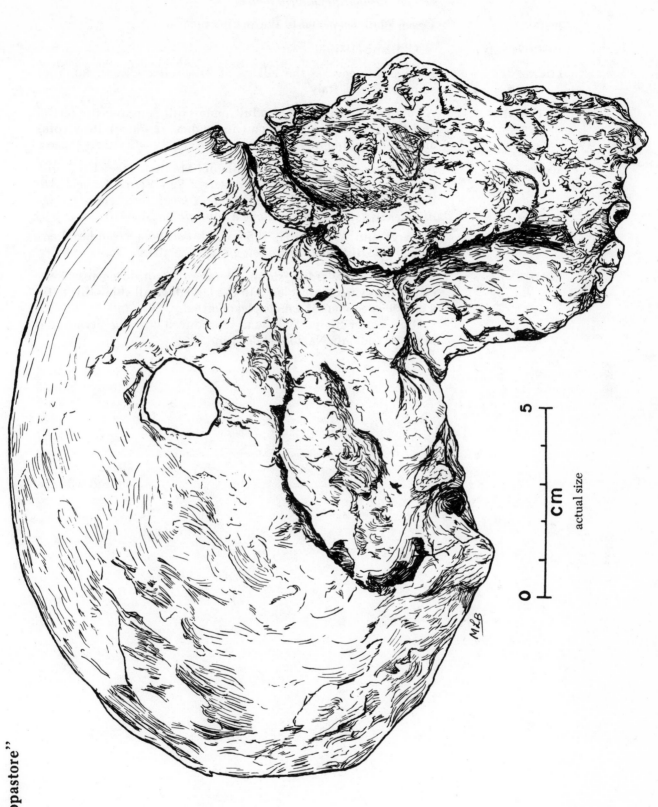

"Saccopastore"

cm
actual size

5

0

113

DESIGNATION:	**"Monte Circeo"** (*Homo sapiens neanderthalensis.*)
DATING:	Upper Pleistocene, early Würm.
DISCOVERED BY:	A. C. Blanc (1939).
LOCALE:	A limestone cave at the village of San Felice Circeo, 80 miles south of Rome, Italy.
MATERIALS:	A complete but encrusted adult male skull (missing the teeth), and the mandible of a second individual found in 1939, plus another mandible in 1950.
SOURCES:	A. C. Blanc, 1939, L'Uomo fossile del Monte Circeo: Un cranio neandertaliano nella Grotta Guattari a San Felice Circeo. *Atti della Reale Accademia Nazionale dei Lincei. Rendiconti, Classe di Scienze fisiche, matematiche e naturali* 29:205–210. S. Sergi and A. Ascenzi, 1955, La Mandibola Neandertaliana Circeo III. *Rivista di Antropologia* 42:337–404.
COMMENT:	The heavy brow ridge, sloping forehead, protruding rear indicating heavy neck muscles, and the large face all show this to be a classic specimen of a male Neanderthal. Drawn after Blanc, 1939, and reproduced with the permission of the Accademia Nazionale dei Lincei.

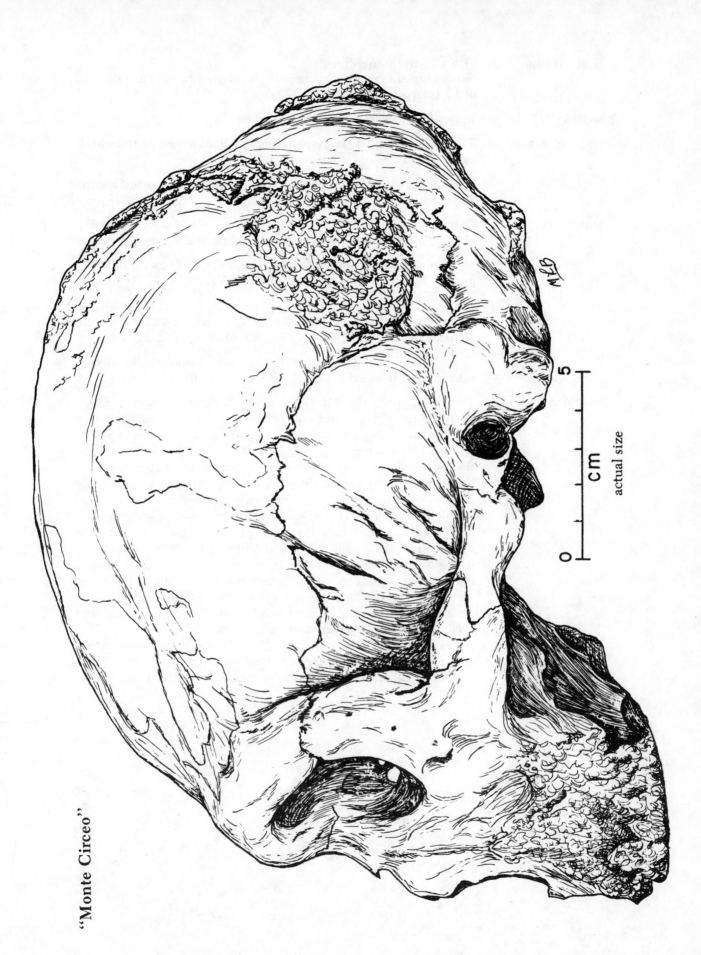

"Monte Circeo"

cm

actual size

5

0

DESIGNATION:	**"La Chapelle-aux-Saints"** (*Homo sapiens neanderthalensis,* often referred to as "The Old Man of La Chapelle-aux-Saints.")
DATING:	Upper Pleistocene, early Würm ?
DISCOVERED BY:	The abbés A. and J. Bouyssonie and L. Bardon, amateur archeologists (1908).
LOCALE:	A small cave near the village of La Chapelle-aux-Saints, Corrèze, southwest France.
MATERIALS:	Burial of a complete skeleton of an aging male. Associated with a Mousterian archeological assemblage (see page 102).
SOURCES:	M. Boule, 1913, *L'homme fossile de La Chapelle-aux-Saints.* Reprinted from *Annales de Paléontologie,* Vols. 6–8. Paris: Masson & Cie, 278 pp.
	W. L. Straus, Jr. and A. J. E. Cave, 1957, Pathology and the posture of Neanderthal Man. *Quarterly Review of Biology* 32:348–363.
	C. L. Brace, 1964, The fate of the "classic" Neanderthals: A consideration of hominid catastrophism. *Current Anthropology* 5:3–43.
COMMENT:	The La Chapelle skeleton is the most complete, the most exhaustively published, the most frequently pictured, and perhaps the most misunderstood Neanderthal specimen. It is an extremely robust male with heavy muscle markings and brow ridges, large face and sloping forehead, and evidently represents an extreme in the Neanderthal range of variation. He had been missing his molars for years prior to death with the consequent reshaping of the jaw and the tooth-bearing part of the face. The result is an atypical Neanderthal and it is unfortunate that the appearance has been taken as characteristic by so many observers. Added to this is the claim made in the original description that he stood in a slouched or stooped fashion. Although there is some indication of arthritis in the spine, there is every reason to regard the posture of this and all other Neanderthals as being fully as erect as that of modern men and even of men more ancient than Neanderthals.
	Drawn from a slide taken by W. W. Howells, and reproduced with the permission of Professor Howells, Harvard University, and Dr. H. V. Vallois, Institut de Paléontologie Humaine, Paris.

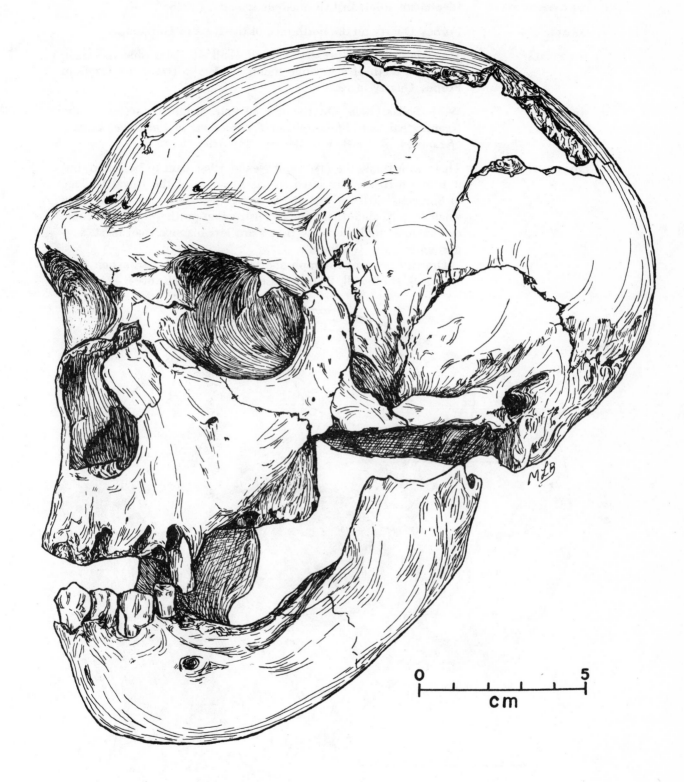

DESIGNATION:	**"Gibraltar"** (*Homo sapiens neanderthalensis.*)
DATING:	Upper Pleistocene ?
DISCOVERED BY:	Lieutenant Flint, British amateur scientist (1848).
LOCALE:	Forbes' Quarry on the north face of the Rock of Gibraltar.
MATERIALS:	The relatively well-preserved adult skull of 1848, plus the skull, jaws, and teeth of a juvenile found at Devil's Tower not far from Forbes' Quarry in 1926.
SOURCES:	W. J. Sollas, 1908, On the cranial and facial characters of the Neanderthal race. *Philosophical Transactions of the Royal Society of London, Series B.* Vol. 199, pp. 281–339.
COMMENT:	This was actually the first adult Neanderthal skull to be found, but it languished unappreciated in the Museum of the Royal College of Surgeons until early in the 20th century. Brain size is relatively small for a Neanderthal (one reconstruction putting it at 1100 cubic centimeters) and the brow ridges are less massive than most, suggesting that this is a female. Drawn from a photograph and reproduced with permission from Dr. D. R. Brothwell, British Museum (Natural History).

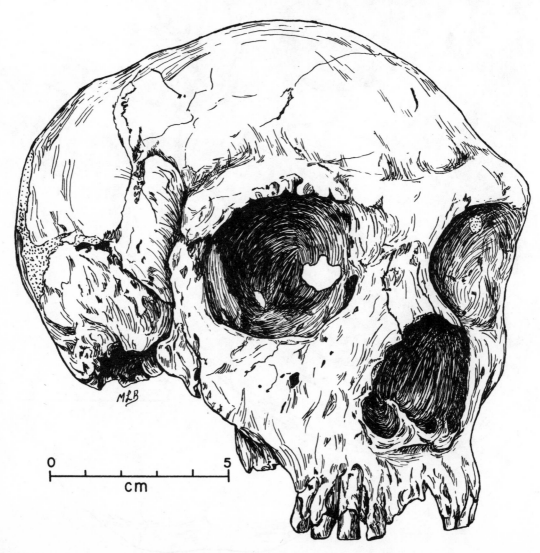

0   5
cm

DESIGNATION: **"Gibraltar"**
(*Homo sapiens neanderthalensis.*)

COMMENT: The key to the difference between Neanderthal and modern form may lie in the large strong teeth which require a large face for support. In many known Neanderthals, the teeth are worn to a phenomenal extent. In spite of their size, the crowns are sometimes completely worn off by early middle age. In the Gibraltar adult this heavy wear is evident. Apparently it had led to the breakdown and loss of the central incisors and some of the molars.

Drawn from a photograph and reproduced with permission from Dr. D. R. Brothwell, British Museum (Natural History).

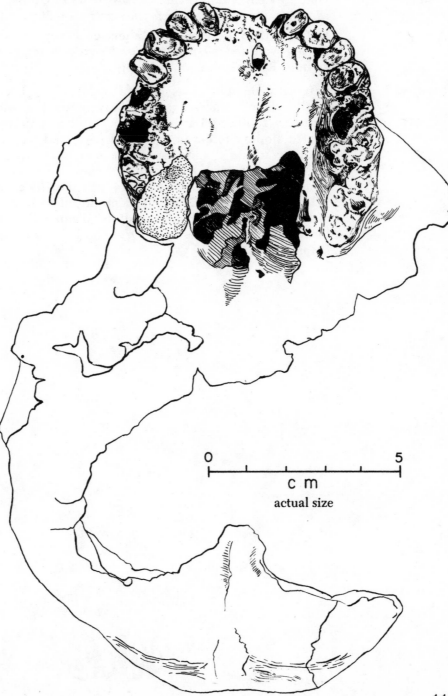

0   5
c m
actual size

DESIGNATION:	**"La Quina"** (*Homo sapiens neanderthalensis.*)
DATING:	Upper Pleistocene, Early Würm.
DISCOVERED BY:	Dr. Henri Martin, French physician and amateur archeologist (1911).
LOCALE:	La Quina, Charente, France (1908–1921).
MATERIALS:	Skull bones, teeth, and other fragments of a number of individuals; parts of the skeleton, skull, jaws, and teeth of a single adult (1911), and the skull, jaws, and teeth of an eight-year-old child (1915). Associated with Mousterian type cultural remains.
SOURCES:	Henri Martin, 1923, L'Homme fossile de la Quina. *Archives de Morphologie générale et expérimentale* 15:1–260. L. Pales, 1958, Les Néanderthaliens en France. In G. H. R. von Koenigswald (ed.), *Hundert Jahre Neanderthaler*, Utrecht, Netherlands: Kemink en Zoon N.V., pp. 32–37.
COMMENT:	Originally thought of as being female largely because of the small mastoid processes and relatively small cranial capacity (just over 1350 cc.), La Quina has a very heavy brow ridge, with related sloping forehead, powerful jaws, and teeth, and a projecting rear end ("bun-shaped" occiput) indicating well-developed neck muscles. As a result of these latter characteristics, some authorities now suspect that it was actually a male. Drawn after Pales, 1958, and reproduced with the permission of Dr. G. H. R. von Koenigswald, the Wenner-Gren Foundation for Anthropological Research, Dr. Léon Pales, and the Musée de l'Homme in Paris.

"La Quina"

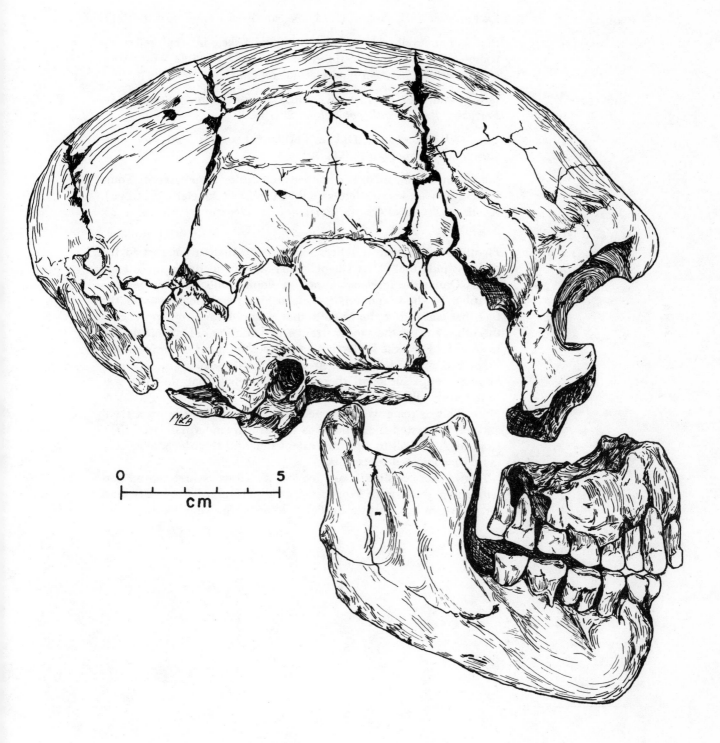

0   5
cm

DESIGNATION:	**"La Ferrassie"** (*Homo sapiens neanderthalensis*)
DATING:	Upper Pleistocene, early Würm.
DISCOVERED BY:	D. Peyrony, French archeologist (1909).
LOCALE:	La Ferrassie rock shelter, Dordogne, southern France (1909–1921).
MATERIALS:	Burial of an adult male and an adult female skeleton, both relatively complete, plus burials of three infants and a possible fetus. Associated with Mousterian type cultural remains.
SOURCES:	L. Capitan and D. Peyrony, 1912, Station préhistorique de La Ferrassie. *Revue Anthropologique* 22:76–99.
	John A. Wallace, 1975, Did La Ferrassie I use his teeth as a tool? *Current Anthropology* 16:393–401.
	Jean-Louis Heim, 1976, Les hommes fossiles de La Ferrassie: Tome I: Le gisement, les squelettes adultes (crâne et squelette du tronc). *Archives de l'Institut de Paléontologie Humaine*, Mémoire 35, 331 pp.
COMMENT:	Pictured is the male, La Ferrassie I. Note that in contrast to the male Neanderthal of La Chapelle-aux-Saints and the possible male of La Quina the forehead is not so sloping and the chin less retreating. In these and other respects, La Ferrassie illustrates the fact that one end of the Neanderthal range of variation approaches the other end of the modern range of variation. The wear on the teeth of La Ferrassie I, like that on the teeth of many other Neanderthals, had proceeded to a degree not usually seen in more recent human groups although there is no evidence that the diet was significantly different from that of the succeeding Upper Palaeolithic. This extreme wear is especially noticeable on the front teeth and has suggested the hypothesis that the Neanderthals used their front teeth as tools to a greater extent than did their technologically more sophisticated successors.

Drawn from a slide taken by W. W. Howells and reproduced with the permission of Professor Howells, Harvard University, and Dr. H. V. Vallois, Institut de Paléontologie Humaine, Paris.

"La Ferrassie"

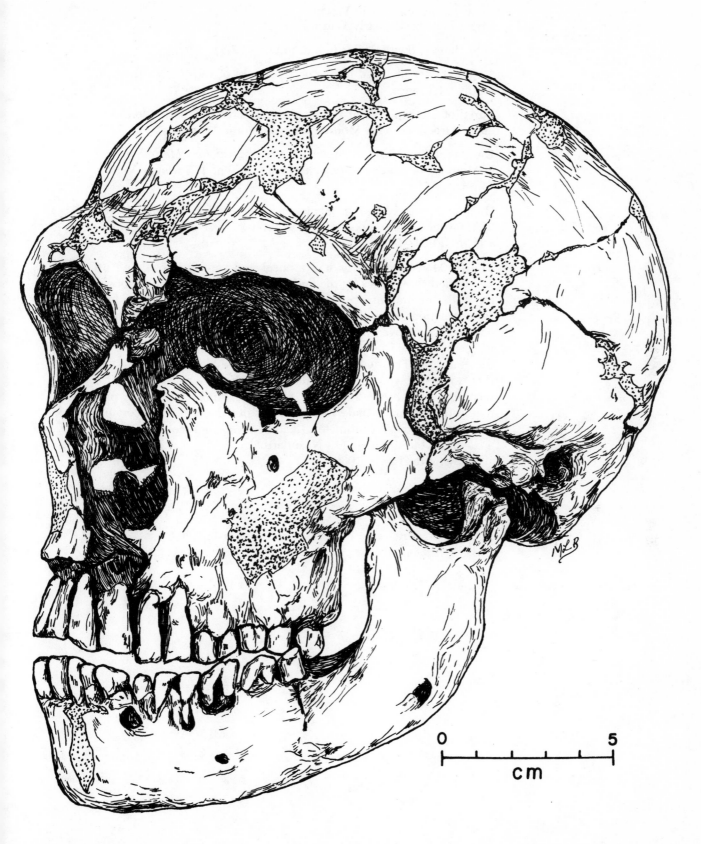

DESIGNATION:	**"Le Moustier"**
	(*Homo sapiens neanderthalensis*, also referred to as "*Homo mousteriensis.*")
DATING:	Upper Pleistocene, early Würm ?
DISCOVERED BY:	Otto Hauser, Swiss dealer in antiquities (1908).
LOCALE:	A rock shelter at Le Moustier on the banks of the Vézère River in southwest France.
MATERIALS:	Burial of a complete but fragmentary skeleton of a late adolescent male. Associated with Mousterian type cultural remains.
SOURCES:	Hans Weinert, 1925, *Der Schädel des eiszeitlichen Menschen von le Moustier in neuer Zusammensetzung.* Berlin: J. Springer Verlag, 55 pp.
	Henrike Hesse, 1966, Zum Schicksal des Neandertales Fundes von le Moustier (*Homo mousteriensis Hauseri*). *Forschungen und Fortschritte* 40:347–348.
COMMENT:	The Le Moustier find has had a checkered career. It was dug up and buried three times by the finder before final excavation. Bought for a Berlin museum by Kaiser Wilhelm, it was subject to four reconstructions, each time losing a few more fragments. Thought to have been destroyed by a bomb in World War II, it was recognized in pieces in 1965 as part of a collection that had been taken by the Russians. The dentition and the battered skull fragments (all that is now known) have been returned to East Germany for a fifth reconstruction. Brow ridges and muscle markings are not as pronounced as in many male Neanderthals, but the large size (cranial capacity 1560 cubic centimeters) and heavy lower face identify it as a classic if youthful male (?) Neanderthal.
	Drawn after Hesse, 1966.

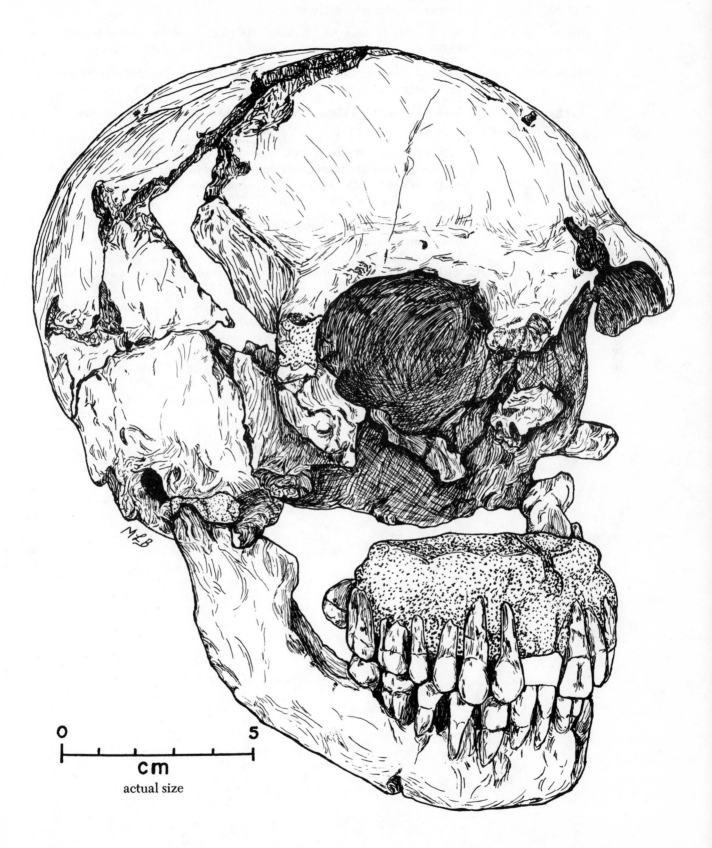

0 — 5
cm
actual size

DESIGNATION:	**"Spy"** (pronounced "Spee") (*Homo sapiens neanderthalensis.*)
DATING:	Upper Pleistocene, early Würm.
DISCOVERED BY:	M. de Puydt and M. Lohest, Belgian amateur archeologists (1886).
LOCALE:	Cave of Bec-aux-Roches in the Commune of Spy, Namur Province, Belgium.
MATERIALS:	Two adult male skeletons buried in a Mousterian culture layer.
SOURCE:	J. Fraipont and M. Lohest, 1887, La race humaine de Néanderthal ou de Canstadt en Belgique. Récherches ethnographiques sur les ossements humains découverts dans les dépôts quaternaires d'une grotte à Spy et détermination de leur âge géologique. *Archives de Biologie* 7:587–757.
COMMENT:	The Spy discoveries show how misleading it can be to base one's impression of a fossil population on only one specimen. Spy I is almost indistinguishable from the original Neanderthal, with the sloping forehead and heavy brow ridge showing the Spy specimen to be a male. In Spy II, the forehead is less sloping and the brow ridge less robust. As a result, some scholars have referred to it as female, and there has been a continuing concern over why the two skulls should look so different. Definitive assignment of sex is not possible, but the two skulls do illustrate something of the range of variation of a Neanderthal population, going from an extreme Neanderthal form in Spy I to a more "modern" appearance in Spy II. Both are drawn from casts.

"Spy II"

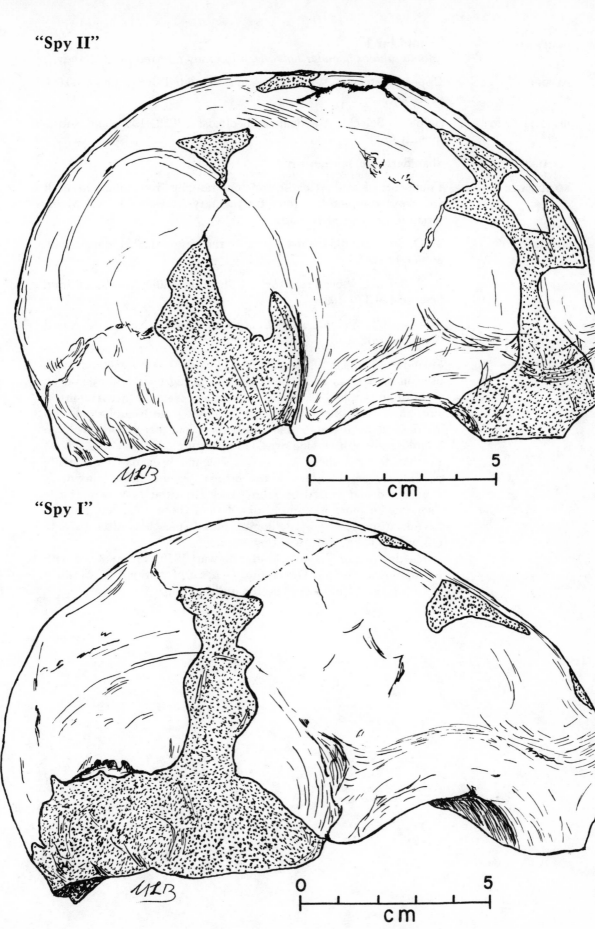

0   cm   5

"Spy I"

0   cm   5

actual size

127

DESIGNATION:	**"Shanidar I"** (*Homo sapiens neanderthalensis,* informally referred to as "Nandy.")
DATING:	Upper Pleistocene, Würm glaciation, ^{14}C dated to just under 50,000 years ago.
DISCOVERED BY:	Ralph S. Solecki, American archeologist (1953, 1957, and subsequently).
LOCALE:	Shanidar cave, northern Iraq.
MATERIALS:	Fragments of at least eight skeletons ranging from infant to adult and from complete to quite fragmentary. Associated with Mousterian archeological remains.
SOURCES:	T. D. Stewart, 1958, First views of the restored Shanidar I skull. *Sumer* 14:90–96.
	R. S. Solecki, 1963, Prehistory in Shanidar Valley, northern Iraq. *Science* 139:179–193.
	———, 1971, *Shanidar: The First Flower People.* New York: Alfred A. Knopf, 302 pp.
COMMENT:	Shanidar was the first Neanderthal site where modern methods were used in excavation and where modern dating techniques could be successfully applied to relatively complete Neanderthal remains *in situ.* Shanidar I, pictured here, is an adult male Neanderthal with all the classic features. This individual had been born with an injured right arm which was later amputated above the elbow. Further, the left cheek and eye socket had been damaged in life possibly causing blindness in the left eye. Several of the individuals were killed and buried by falling rock from the cave roof. Hence they can be more accurately dated than those Neanderthals that have been preserved by deliberate burial since it is often hard to tell what level a grave was dug *from.* The drawing of Shanidar I, after Stewart 1958, is reproduced with the permission of the Directorate General of Antiquities, Ministry of Culture and Guidance, The Republic of Iraq.

128

"Shanidar I"

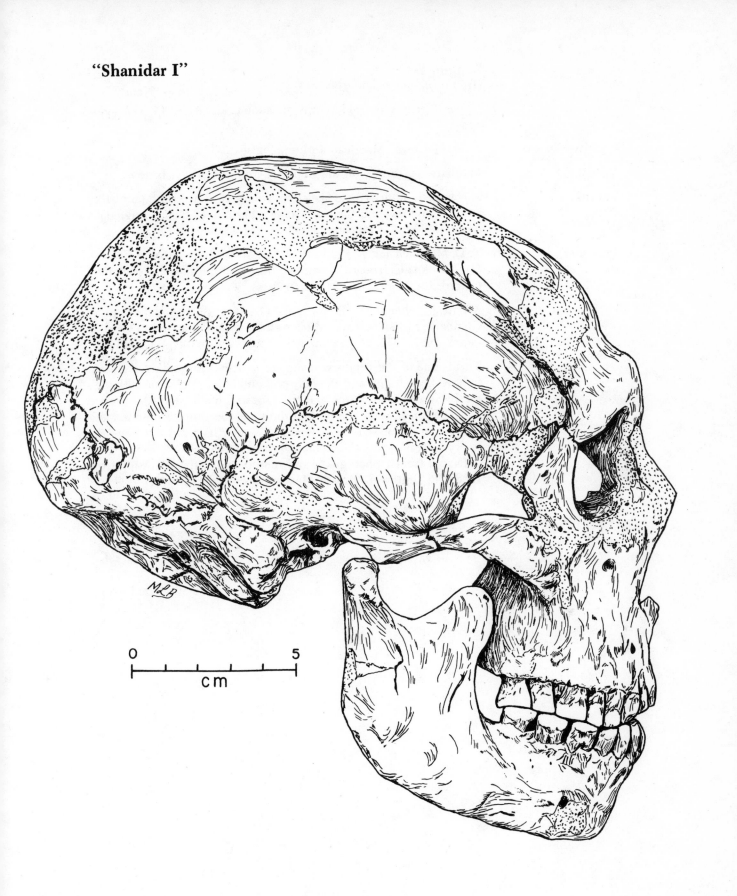

0   5
cm

DESIGNATION:	**"Tabūn I"** (*Homo sapiens neanderthalensis.*)
DATING:	Upper Pleistocene, mid-Würm glaciation, about 50,000 (?) years ago.
DISCOVERED BY:	T. D. McCown, American anthropologist (1932).
LOCALE:	Mugharet-et Tabūn, on the slopes of Mount Carmel, Israel.
MATERIALS:	A complete adult female skeleton. Tabūn II is a robust (probably male) mandible, and there are some other miscellaneous fragments. Associated with a Mousterian cultural assemblage.
SOURCES:	T. D. McCown and A. Keith, 1939, *The Stone Age of Mount Carmel, II. The Fossil Human Remains from the Levalloiso-Mousterian.* Oxford, England: The Clarendon Press, 390 pp.
	Arthur J. Jelinek, W. R. Farrand, G. Haas, A. Horowitz, and P. Goldberg, 1973, New excavations at the Tabūn Cave, Mount Carmel, Israel, 1967–1972: A preliminary report. *Paleorient* 1:151–183.
COMMENT:	This is one of the very few complete female Neanderthal skeletons to have been found and the only one published. Note that although the brow ridges are large and the face and teeth protrude in good Neanderthal fashion, the back of the skull (occiput) is rounded and modern in form. Evidently female Neanderthals did not have the heavy neck muscles which characterized the males.
	Drawn from a photograph and reproduced with permission from Dr. D. R. Brothwell, British Museum (Natural History).

"Tabūn I"

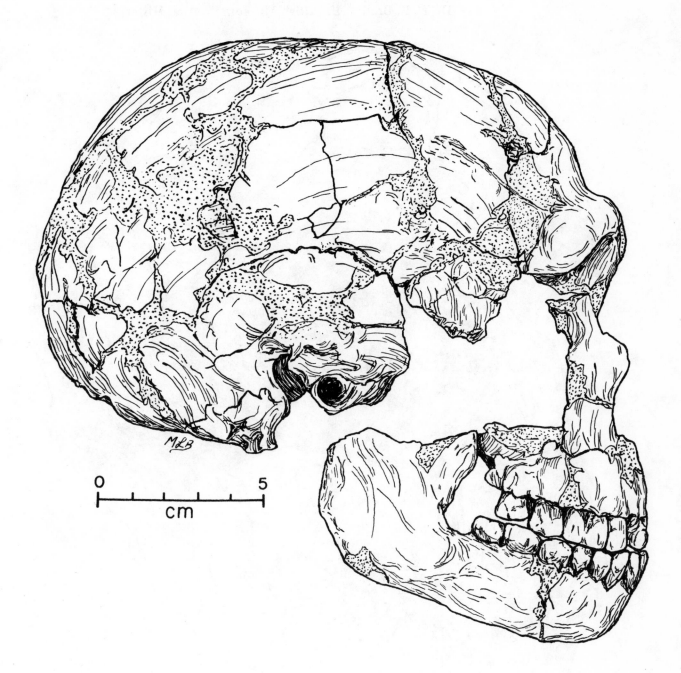

0 ———————————— 5
cm

DESIGNATION: **"Tabūn I"**
(*Homo sapiens neanderthalensis.*)

COMMENT: This view is particularly effective in showing the heavy brow ridge development and the relatively large dentition, especially the incisor teeth.

Drawn from a photograph and reproduced with permission of Dr. D. R. Brothwell, British Museum (Natural History).

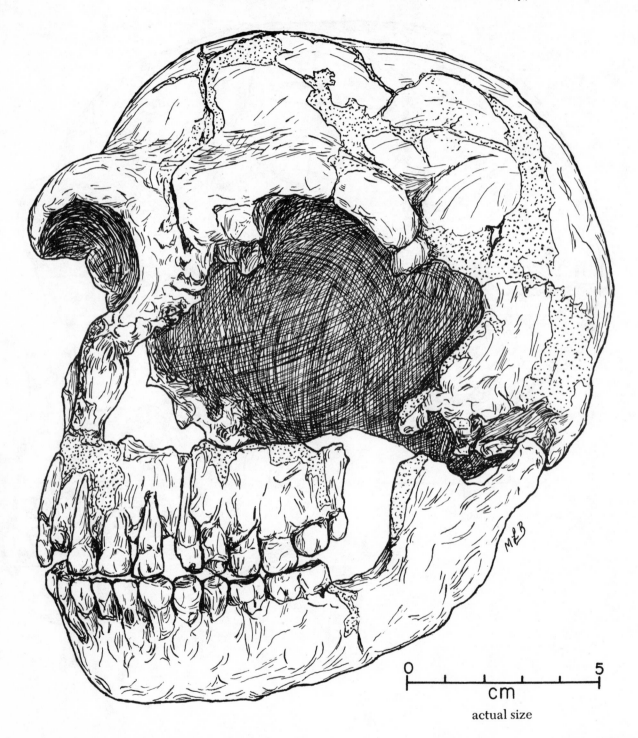

0   5
cm
actual size

DESIGNATION: **"Mapa"**
(Homo sapiens neanderthalensis.)

DATING: Early Upper Pleistocene?

DISCOVERED BY: Farmers digging fertilizer and Ju-kang Woo, Chinese anthropologist (1958).

LOCALE: A limestone cave at Mapa village, Kwangtung Province north of Canton in China.

MATERIALS: A skull cap with the right eye socket included.

SOURCES: Ju-kang Woo and Ru-ce Peng, 1959, Fossil human skull of early paleoanthropic stage found at Mapa, Shaoquan, Kwangtung Province. *Vertebrata Palasiatica* 3:176–182.

Kwang-chih Chang, 1962. New evidence on fossil man in China. *Science* 136:749–760.

COMMENT: The date may be late Middle or early Upper Pleistocene, but there can be no doubt that this is a Chinese representative of the Neanderthal stage.
Drawn after Woo and Peng, 1959.

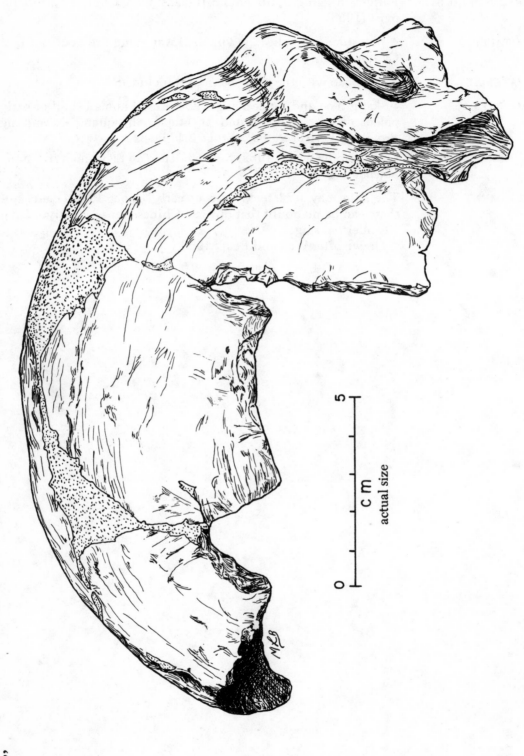

"Mapa"

134

MODERN HOMO SAPIENS

The distinction between Neanderthal and modern is arbitrary, and since we take the position that modern form evolved from Neanderthal form, obviously there will be specimens, dating from the time during which the change took place, that will be difficult to put in either category. The term "Neanderthaloid" has been used to designate those forms which are neither clearly Neanderthal nor clearly modern, that is, Amud, Shkūl V, Jebel Irhoud, and possibly others. Even this is an arbitrary matter of judgment since many early moderns had one or another feature clearly recalling a Neanderthal ancestry while all of the remaining features might have been modern.

Evidently what we regard as modern form developed gradually over a span of 20,000 to 30,000 years, nor have these trends yet stopped. Still further reductions in muscularity and tooth and face robustness have taken place in the last several thousand years, and there is every evidence that they are continuing at present.

The cultural remains associated with the early moderns, the Upper Palaeolithic tool-making traditions, developed out of Mousterian predecessors in just those parts of the Old World where the cold-adapted Mousterian had previously prevailed. The Upper Palaeolithic represented a more efficient cultural adaptation to cold climatic conditions and enabled its early modern possessors to spread north and, especially, east into previously unoccupied territory—ultimately leading to the spread across the Bering land bridge into the New World.

MODERN SPREAD

(The sites are too numerous to be plotted)

■ Core area in which the Upper Palaeolithic developed.

⟶ Actual spread of Upper Palaeolithic populations.

---⟶ Diffusion of certain Upper Palaeolithic tool-making
techniques adopted by previously existing populations.

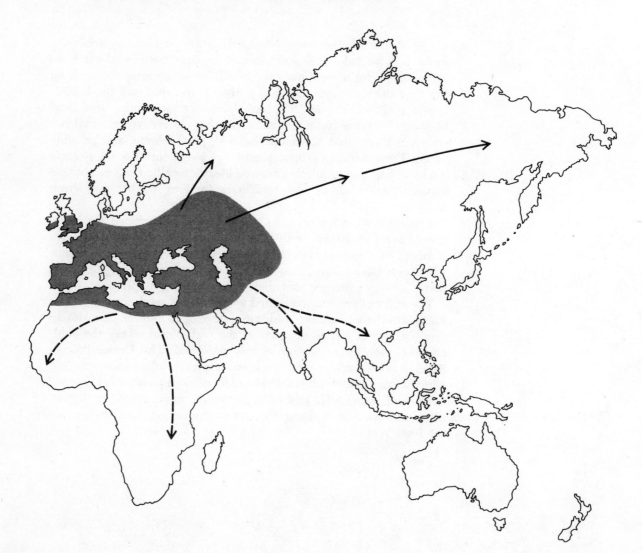

UPPER PALAEOLITHIC TOOLS

FLINT TOOLS: *a, b*) "Knife points"; *c*) Truncated blade; *d, e, f, g*) Gravers or burins; *i, j, l, o*) Scrapers; *h*) Strangulated blade; *k*) Piercer; *m*) Blade core; *n*) Miscellaneous notched piece.

Associating the above with specific Upper Palaeolithic traditions: *a*) Chatelperronian; *b*) Gravettian; *e, f, h, i*) Aurignacian; *d*) Perigordian; *k*) Solutrean; *g, l, m, n, o*) Magdalenian. The complexity of tool types seen in the Upper Palaeolithic represents a refinement of the initial diversification seen in the Mousterian. Worked bone of increasing sophistication also appears in the Upper Palaeolithic.

(Drawings reprinted from Oakley 1950, by permission of the Trustees of the British Museum [Natural History].)

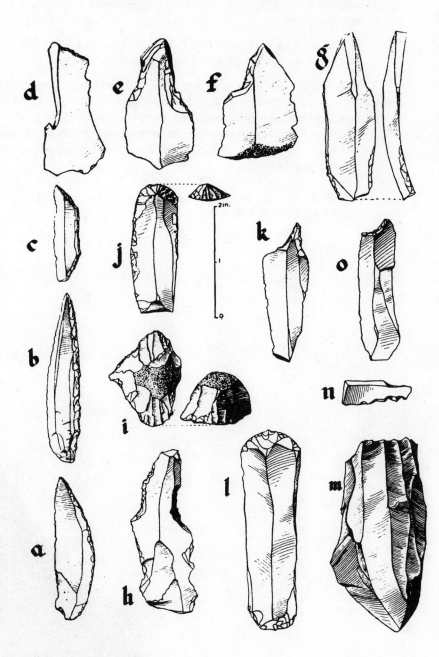

DESIGNATION:	**"Amud"** (*Homo sapiens neanderthalensis* ?)
DATING:	Upper Pleistocene.
DISCOVERED BY:	Hisashi Suzuki, Japanese anthropologist (1961).
LOCALE:	Amud cave in Wadi Amud near Galilee, Israel.
MATERIALS:	A relatively complete adult male skeleton. Associated with archeological remains that are transitional between the Mousterian (or Middle Palaeolithic) and the Upper Palaeolithic.
SOURCE:	Hisashi Suzuki, 1965, A Palaeoanthropic man from the Amud Cave, Israel (preliminary report). *Communications to VII International Congress of Anthropological and Ethnological Sciences, Moscow, August 3–10, 1964.* pp. 1–8. Tokyo University Expedition to Western Asia, Tokyo.
	H. Suzuki and F. Takai (eds.), 1970, *The Amud Man and His Cave Site.* Tokyo: Keigaku Publishing Company, 530 pp.
COMMENT:	Brow ridges are somewhat less robust than those of fully developed male Neanderthals and the dentition and tooth-bearing part of the face are sufficiently reduced so that a definite chin is visible. All told, the form borders on the primitive modern in much the same fashion as does that of the Skhūl specimens from Mount Carmel, that is, just half way between fully Neanderthal and fully modern form.
	Drawing made with permission from a photograph kindly furnished by Professor Suzuki.

138

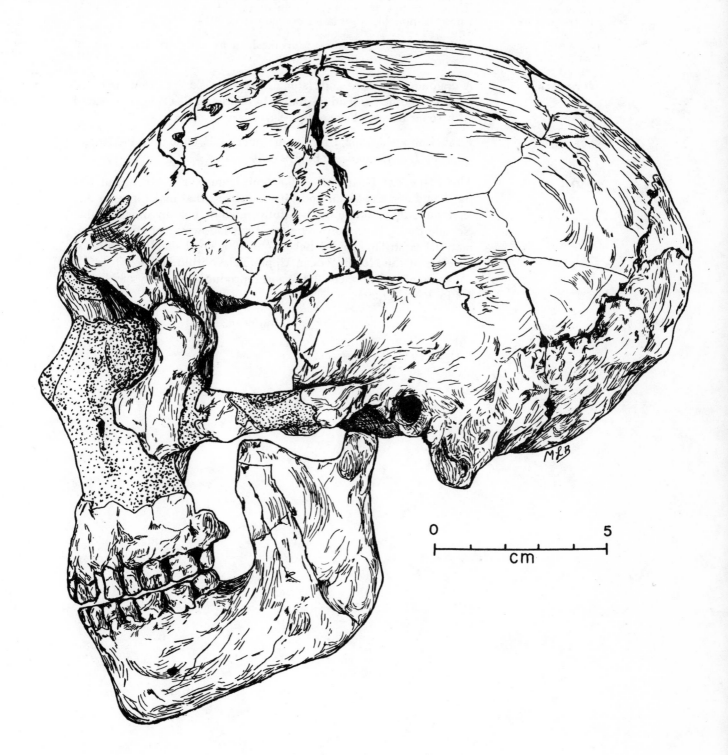

DESIGNATION:	**"Jebel Irhoud"** (sometimes spelled "Jebel Ighoud") (*Homo sapiens neanderthalensis* ?)
DATING:	Upper Pleistocene, early Würm glaciation?
DISCOVERED BY:	Emile Ennouchi, Algerian archeologist (1961).
LOCALE:	In a mine not far from Marrakesh near the Atlantic coast of Morocco.
MATERIALS:	A cranium with face and, later (1962), the skull cap minus facial parts of another individual found by Carleton S. Coon. Associated with some Middle Palaeolithic flakes.
SOURCES:	Emile Ennouchi, 1962, Un Néandertalien: L'homme du Jebel Irhoud (Maroc). *L'Anthropologie* 66:279–299.
COMMENT:	This is the first Neanderthal-like skull from North Africa, but like the Amud and the Skhūl finds, it has aspects that make it seem like an intermediate form, neither fully modern nor an unmistakable Neanderthal. The morphological foreshadowing of the Afalou material is striking and Jebel Irhoud is a good candidate for the ancestor of the North African Upper Palaeolithic people. Drawn after Ennouchi, 1962, and reproduced courtesy of Masson & Cie, Paris, France.

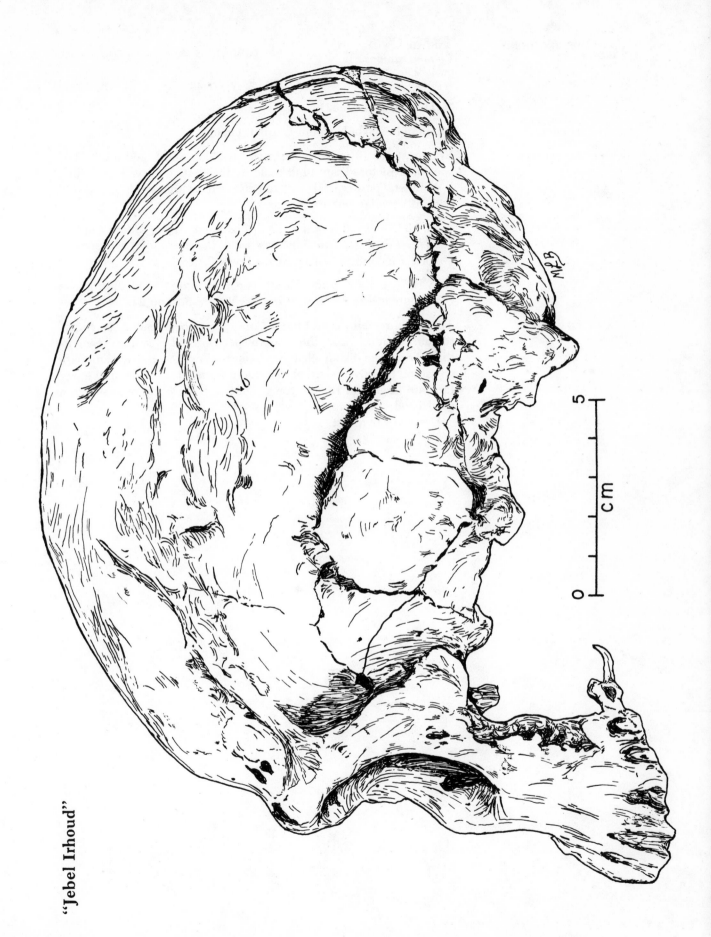

"Jebel Irhoud"

DESIGNATION:	**"Skhūl V"** (*Homo sapiens* ?)
DATING:	Upper Pleistocene, about 35,000 (?) years ago.
DISCOVERED BY:	T. D. McCown (1932).
LOCALE:	Mugharet es-Skhūl, Mount Carmel, Israel.
MATERIALS:	Fragments of 10 individuals: male, female, infant, and adult. The specimen pictured, an adult male, is the best preserved, including a complete post-cranial skeleton as well as the skull. Associated with Mousterian cultural remains.
SOURCES:	T. D. McCown and A. Keith, 1939, *The Stone Age of Mount Carmel, II. The Fossil Human Remains from the Levalloiso-Mousterian.* Oxford, England: The Clarendon Press, 390 pp.
	E. S. Higgs, 1961, Some Pleistocene faunas of the Mediterranean coastal areas. *Proceedings of the Prehistoric Society* 27:144–154.
COMMENT:	The Skhūl remains from Mount Carmel have been called "Neanderthal*oid*," that is, within the modern spectrum but recalling Neanderthal form. The Skhūl Neanderthal*oids* illustrate the evolutionary transition from Neanderthal to modern form. Brow ridges and lower face are large and rugged for a modern, but clearly reduced from the Neanderthals proper. Drawn from a cast.

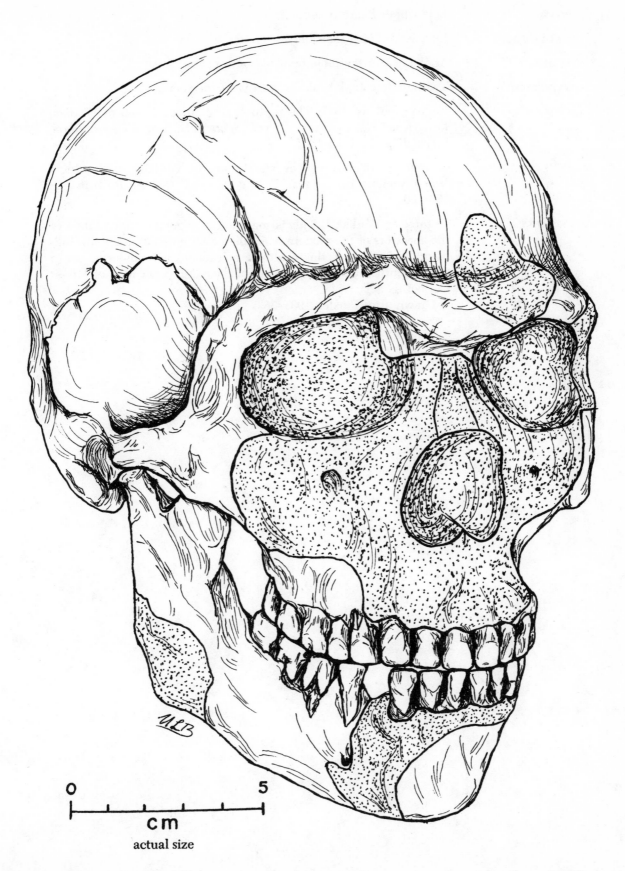

0   5

cm

actual size

DESIGNATION:	**"Wadjak I"**
	(*Homo sapiens.*)
DATING:	Late Upper Pleistocene ?
DISCOVERED BY:	Eugene Dubois (1889 and 1890).
LOCALE:	At Wadjak in the mountains of south central Java.
MATERIALS:	Two nearly complete skulls with jaws and teeth.
SOURCES:	Eugene Dubois, 1921, The proto-Australian fossil man of Wadjak, Java. *Koninklijke Akademie van Wetenschappen te Amsterdam, Proceedings* 23:1013–1051.
	G. Pinkley, 1935–1936, The significance of Wadjak man: a fossil *Homo sapiens* from Java. *Peking Natural History Bulletin* 10: 183–200.
COMMENT:	Pictured is Wadjak I. This is regarded as a good representative of the ancestors of the modern Australian aborigines. While not a full-blown Neanderthal, this could be considered a Neanderthal*oid*, a sort of southeast Asian equivalent of the Skhūl population of Mount Carmel, Israel.
	Drawn after Pinkley 1935–36.

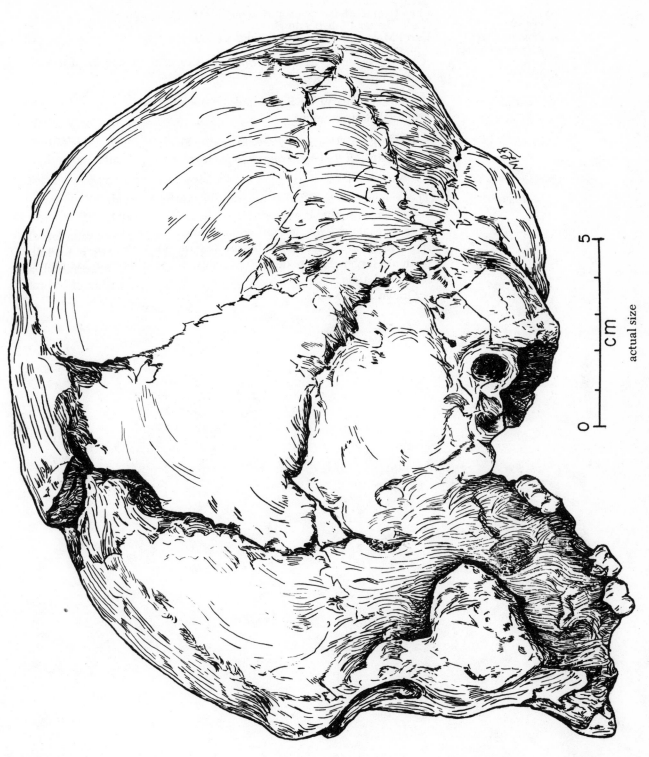

"Wadjak I"

DESIGNATION:	**"Předmost III"** (*Homo sapiens.*)
DATING:	Late Upper Pleistocene, just over 30,000 years ago.
DISCOVERED BY:	Karel Maška, Czech anthropologist (1894).
LOCALE:	A collective grave in a large open air site near Prerov in Moravia, central Czechoslovakia.
MATERIALS:	Remains of nearly 30 individuals. Associated with Upper Palaeolithic cultural remains.
SOURCE:	J. Matiegka, 1934, *Homo Předmostensis, Fosilní Člověk Z Předmostí Na Moravě. I. Lebky.* Prague: Nákladem České Akademie Věd A Umění, 145 pp.
COMMENT:	The specimens found are among the earliest "moderns" known. Some, as with the Předmost III male skull pictured, look almost Neanderthal*oid*. One has an even heavier brow ridge but others show relatively smooth foreheads. On the average, the jaws and teeth are larger than modern Europeans, but smaller than the Skhūl Neanderthal*oids*. Drawn from a cast.

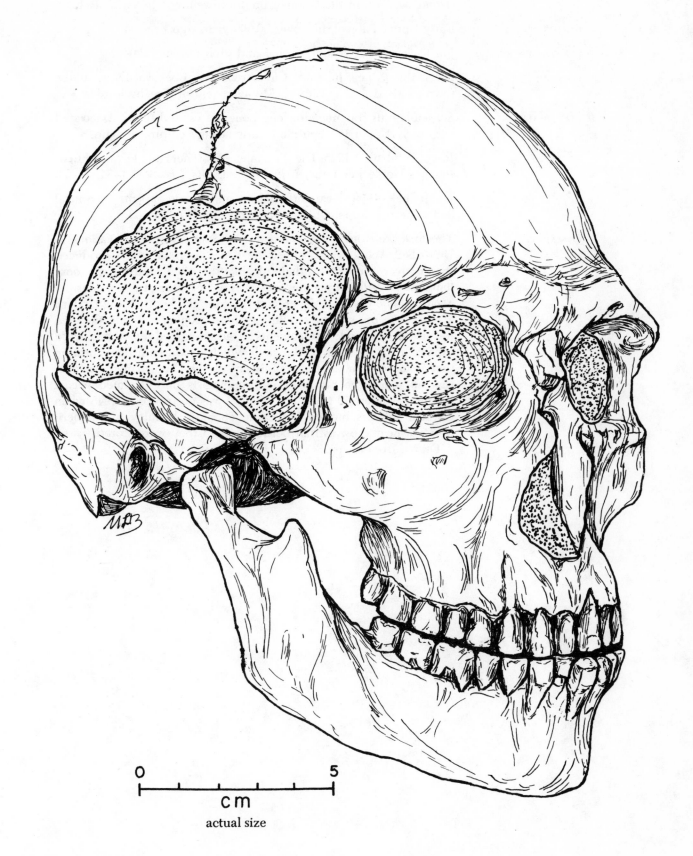

0   5

cm

actual size

DESIGNATION:	**"Mladeč V"** (*Homo sapiens*, in the German literature referred to as Lautsch.)
DATING:	Early Upper Palaeolithic, over 30,000 years ago.
DISCOVERED BY:	Jan Knies, Czech school teacher and prehistorian (1904).
LOCALE:	Originally Prince Johann's Cave, renamed Bočkova Díra, in the town of Mladeč near Litovel, Moravia, central Czechoslovakia.
MATERIALS:	Six skulls plus fragments of long bones, jaws, and teeth. Associated with early Upper Palaeolithic (Aurignacian) cultural remains.
SOURCES:	Josef Szombathy, 1925, Die diluvialen Menschenreste aus der Fürst-Johanns-Höhle bei Lautsch in Mähren. *Die Eiszeit* 2:1–34, 73–95. Jan Jelínek, 1969, Neanderthal man and *Homo sapiens* in central and eastern Europe. *Current Anthropology* 10:475–503.
COMMENT:	The first skeletons were found during Szombathy's excavations in 1881–1882. Work was resumed by Knies in 1903 which produced the discoveries of 1904. Mladeč V, like the more robust of the contemporary Předmost people, preserves clear traces of a Neanderthal ancestry. The bones of the cranial vault are thicker than the modern average, the brow ridge swelling is more pronounced, and the area for the neck-muscle attachments is accompanied by a backward expansion of the rear of the skull which produces a configuration that is more than faintly reminiscent of the "bun-shaped occiput" that an older generation of anthropologists once claimed was a unique Neanderthal trait. The original was destroyed in 1945 when German soldiers burned the castle (Schloss Mikulov in southern Moravia near the Austrian border) where they were stored. Drawn from a cast.

148

"Mladěc V"

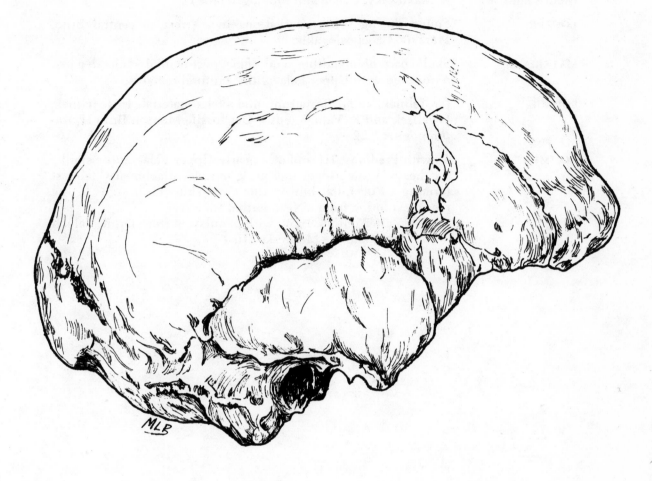

0 _____ 5
cm

149

DESIGNATION:	**"Brno II"**
	Brünn II (*Homo sapiens.*)
DATING:	Upper Pleistocene, late Würm ?
DISCOVERED BY:	A. Makowsky, Czech anthropologist (1891).
LOCALE:	Found during sewer excavations in a street in central Brno, Moravia, Czechoslovakia.
MATERIALS:	Skull, part of mandible, and some post-cranial skeletal parts. Associated with Upper Palaeolithic cultural remains.
SOURCE:	Jan Jelinek, 1959, Bestattung und skelettmaterial. In J. Jelinek, J. Palíšek and K. Valoch (eds.), Der fossile Mensch Brno II, *Anthropos* 9:17–22.
COMMENT:	As with Předmost III and other early Upper Palaeolithic skulls, the heavy brow ridge and neck muscle attachments (almost forming an occipital bun in this case) show clear signs of the recent evolution from a Neanderthal ancestry.
	Drawn after Jelinek 1959, with permission from Dr. K. Valoch of *Anthropos* and the Moravské Museum.

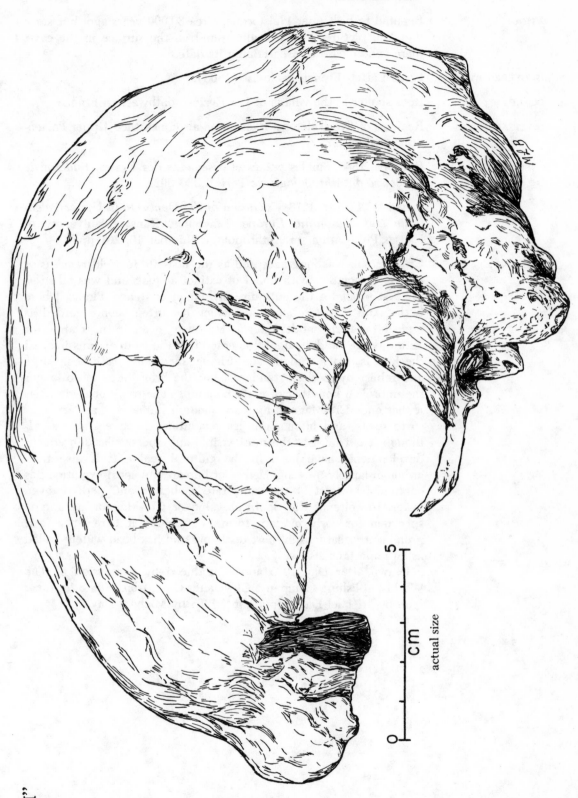

"Brno II"

5

cm

actual size

0

151

DESIGNATION:	**"Cro-Magnon"** (*Homo sapiens.*)
DATING:	Reputed to be Upper Pleistocene, circa 30,000 years ago, but since the skull pictured was actually found on the surface in the cave, there is no means of ascertaining its date.
DISCOVERED BY:	Louis Lartet, French geologist (1868).
LOCALE:	Rock shelter at the village of Les Eyzies, southwestern France.
MATERIALS:	Remains of five skeletons. Excavation also unearthed Upper Palaeolithic artifacts.
SOURCES:	P. Broca, 1868, Sur les crânes et ossements des Eyzies. *Bulletin de la Société d'Anthropologie de Paris* 3:350–392.
	David W. Frayer, 1978, *Evolution of the Dentition in Upper Paleolithic and Mesolithic Europe.* Lawrence, Kansas: University of Kansas Publications in Anthropology Number 10, 201 pp.
COMMENT:	Cro-Magnon was long accepted as the first site to yield remains of men in association with bones of extinct animals and was taken as the first proof for the antiquity of human existence. Pictured here is the skull of the robust male which has often been regarded as typical for early modern form. However, even if the claims for antiquity are valid, it should be noted that the teeth are missing and the face is badly eroded. The original face must have been large and robust by modern standards, but the brow ridge is no longer the unbroken bony bar of early hominids. The traditional view has it that once Cro-Magnon appears, modern humanity has "arrived," so to speak, and human physical evolution effectively ceases. Although it is true that the most visible and spectacular subsequent developments have been in the cultural realm, it is also true, although largely unrecognized, that human biological evolution has continued at least at the same rate as before and perhaps even faster. In spite of the doubtful value of Cro-Magnon as a type specimen for early modern form, we include it here since it is prominently featured in most textbooks and has been widely hailed as an important discovery.

Drawn after Day 1965 and reproduced by permission of *The World Publishing Company*, Cleveland, Ohio, from *Guide to Fossil Man* by Michael Day, copyright © 1965 by Michael Day.

"Cro-Magnon"

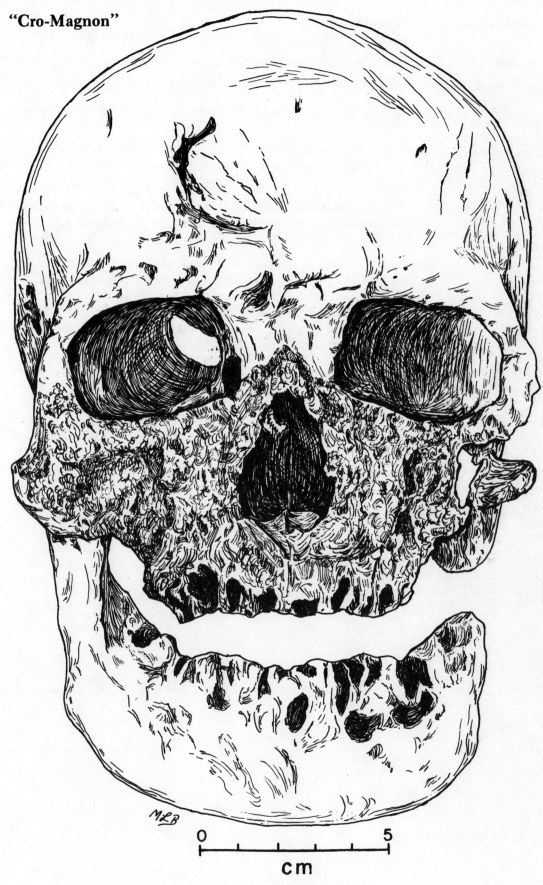

0   5

cm

actual size

153

DESIGNATION:	**"Combe Capelle"** (*Homo sapiens*, once erroneously called *"Homo aurignacensis"*)
DATING:	Late Upper Pleistocene, just over 30,000 years ago.
DISCOVERED BY:	Workmen hired by Otto Hauser, Swiss dealer in antiquities (1909).
LOCALE:	Rock shelter near Montferrand in the Dordogne region of southwest France.
MATERIALS:	Burial of a complete and relatively well-preserved adult male skeleton. Associated with early Aurignacian cultural remains.
SOURCE:	H. Klaatsch and O. Hauser, 1910, Homo Aurignacensis Hauseri, ein paläolithischer Skelettfund aus dem unteren Aurignacien der Station Combe-Capelle bei Montferrand (Périgord). *Prähistorische Zeitschrift* 1:273–338.
COMMENT:	As with other early modern skeletal material, the tooth-bearing parts of the face and jaws are distinctly larger than is typical for modern Europeans, although reduced from fully Neanderthal size. Sold by Hauser along with Le Moustier, it was exhibited in the Staatlichen Museum für Vor und Frühgeschichte in Berlin. It was destroyed when the museum was the recipient of a direct hit by an American "blockbuster" bomb in an air raid in 1945. Drawn from a cast.

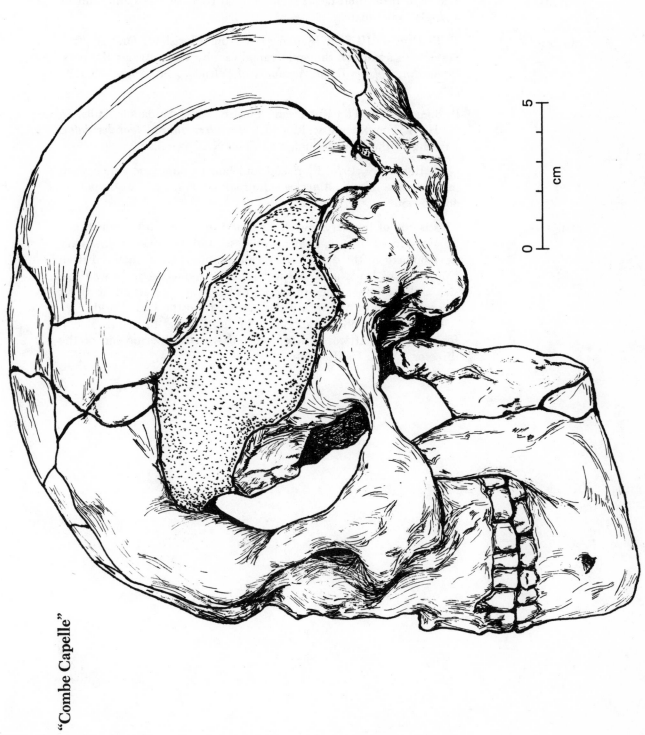

"Combe Capelle"

155

DESIGNATION:	**"Florisbad"** (*Homo sapiens.*)
DATING:	Upper Pleistocene, late Würm possibly as old as 33,000 B.C. and in the minds of some, maybe even older than 40,000 B.C.
DISCOVERED BY:	T. F. Dreyer, South African anthropologist (1932).
LOCALE:	A spring deposit, Florisbad, north of Bloemfontein in the Transvaal, South Africa.
MATERIALS:	The right orbit, mid-facial skeleton, and parts of the skull vault of a single individual.
SOURCES:	T. F. Dreyer, 1935, A human skull from Florisbad, Orange Free State, with a note on the endocranial cast by C. U. Ariëns Kappers. *Proceedings of the Royal Academy of Science, Amsterdam.* 38:119– 128.
	R. Singer, 1958, The Rhodesian, Florisbad and Saldanha skulls. In G. H. R. von Koenigswald (ed.), *Hundert Jahre Neanderthaler.* Utrecht, Netherlands: Kemink en Zoon N.V., pp. 52–62.
	G. P. Rightmire, 1978, Florisbad and human population succession in southern Africa. *American Journal of Physical Anthropology* 48:475–486.
COMMENT:	This is one of the very few skulls found in Africa so far which can be considered the equivalent in date and relative evolutionary development to the early Upper Palaeolithic skeletal material from Europe. This leads us to suspect that the reduction in brow ridge size and the appearance of modern face form proceeded at about the same pace in Africa as it did elsewhere in the Old World.
	Drawn from a photograph taken in the National Museum, Bloemfontein, South Africa, and reproduced with the permission of the director, Dr. A. C. Hoffman.

"Florisbad"

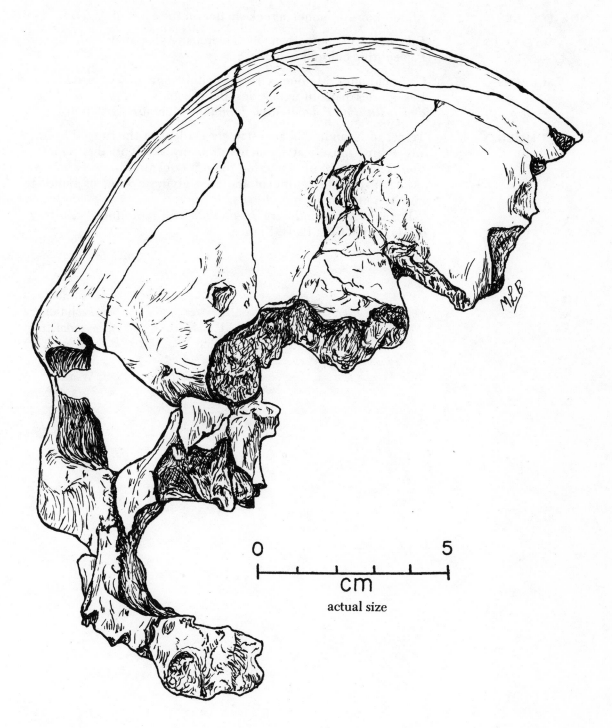

0   5
|__|__|__|__|__|__|__|__|__|__|
         cm
      actual size

DESIGNATION:	**"Afalou"**
	Mouillian men or Mechta-el-Arbi men (*Homo sapiens.*)
DATING:	Upper Pleistocene.
DISCOVERED BY:	C. Arambourg (1928).
LOCALE:	Afalou-bou-Rhummel, a rock shelter on the coast of Algeria.
MATERIALS:	Fragments of more than 50 skeletons. Associated with Upper Palaeolithic cultural remains.
SOURCES:	C. Arambourg, M. Boule, H. V. Vallois, and R. Verneau, 1934, Les grottes paléolithiques des Béni-Segoual, Algérie. *Archives de l'Institut de Paléontologie humaine.* Mémoire 13, 239 pp.
COMMENTS:	Traces of Neanderthal ancestry are evident in the rugged brows and robust muscle attachments of many skulls in this population. The custom of extracting the upper central incisors has resulted in the shortening of the face giving a kind of bulldog impression to many individuals.
	Drawn after Arambourg *et al.* 1934 and reproduced courtesy of Masson & Cie., Paris, France.

"Afalou"

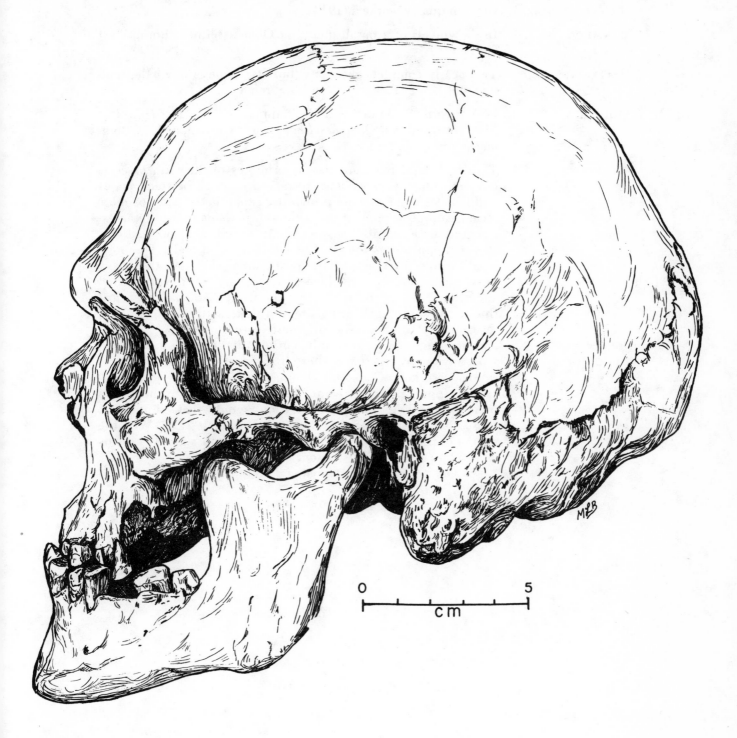

DESIGNATION:	**"Obercassel"**
	(*Homo sapiens.*)
DATING:	Late Upper Pleistocene, late Würm.
DISCOVERED BY:	German quarry workers (1914).
LOCALE:	In the gravels of a basalt quarry at Obercassel near Bonn, West Germany.
MATERIALS:	Two nearly complete skeletons (burials), male and female, with traces of Upper Palaeolithic cultural remains.
SOURCE:	M. Verworn, R. Bonnet and G. Steinmann, 1919, *Der Diluviale Menschenfund von Obercassel bei Bonn.* Wiesbaden: J. F. Bergmann Verlag, 193 pp.
COMMENT:	The Obercassel finds are classic examples of the form of early modern *Homo sapiens.* The average difference between males and females appears more pronounced than is the case for more recent European populations although perhaps not quite as extreme as was true for the preceding Neanderthal populations. Pictured opposite is the skull of a middle-aged male. The powerful muscle markings on the post-cranial skeleton, the heavy brow ridges, and the large face all clearly point to a Neanderthal ancestry. It should be noted that the front teeth had been lost through heavy usage during life and that, with the consequent resorbtion of the tooth-bearing part of the upper jaw, the face appears much shorter than it had originally been. As it is drawn, it is obviously much too short in proportion to its width. A proper restoration, opening the jaw and inserting the long-missing upper incisors, would lengthen the face by nearly 20 percent. The length-width proportions would then be close to those visible in the facial skeleton of the female skull. With the available illustrations, however, it was not possible to do this without unacceptable distortion.
	Drawn after Verworn, Bonnet and Steinmann 1919.

160

"Obercassel," male

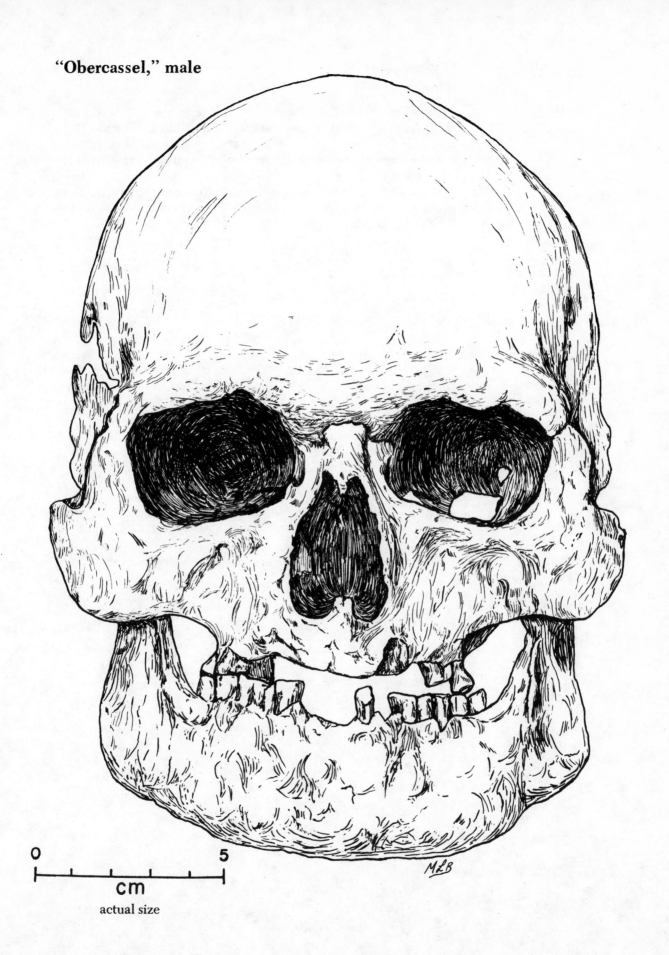

0   5

cm

actual size

M£B

DESIGNATION: **"Obercassel"**
(*Homo sapiens.*)

COMMENT: Pictured is the skull of the young adult female from Obercassel. While far less massive than the male, the facial skeleton is large and well developed, being comparable to the masculine form in the spectrum of the range of variation of twentieth century European populations.

Drawn from Verworn, Bonnet and Steinmann 1919, and reproduced with the permission of J. F. Bergmann, Verlagsbuchhandlung, Munich.

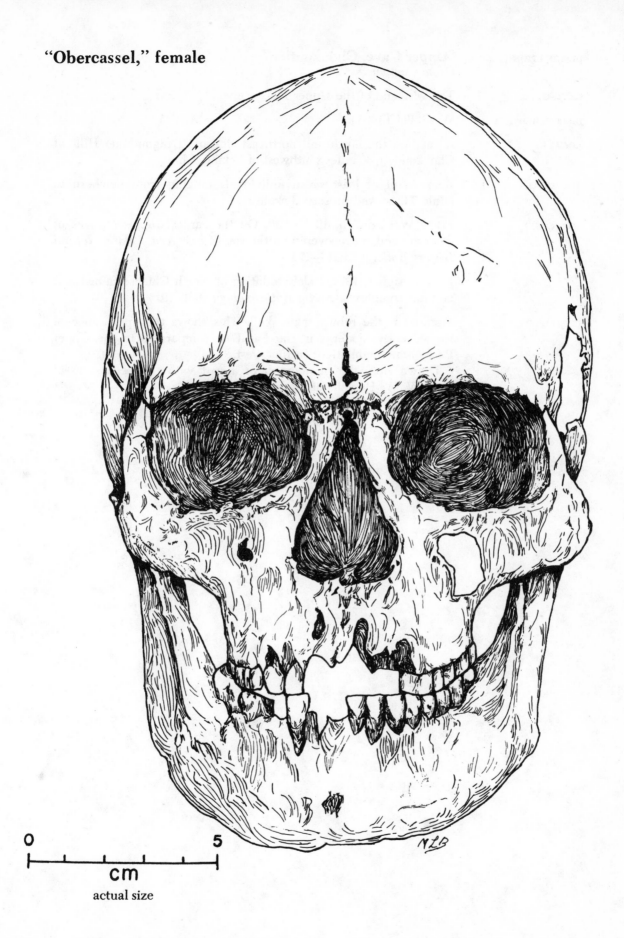

"Obercassel," female

0   5

cm

actual size

DESIGNATION:	**"Upper Cave, Choukoutien"** (*Homo sapiens.*)
DATING:	The very end of the Upper Pleistocene.
DISCOVERED BY:	W. C. Pei (1933).
LOCALE:	A cave on the north and northeast side of "Dragon-bone Hill" at Choukoutien, 30 miles southwest of Peking, China.
MATERIALS:	Remains of at least seven individuals ranging from newborn to adult. Three well preserved skulls.
SOURCES:	Franz Weidenreich, 1938–1939, On the earliest representatives of modern mankind recovered on the soil of east Asia. *Peking Natural History Bulletin* 13:161–174. Jean S. Aigner, 1972, Relative dating of North Chinese faunal and cultural complexes. *Arctic Anthropology* 9:36–79.
COMMENT:	Pictured is the robust male skull. This shows the appearance of rugged early moderns in the Far East comparable to the Upper Palaeolithic populations further west. Drawn from the photograph of a cast with the permission of Prof. W. W. Howells of Harvard University.

164

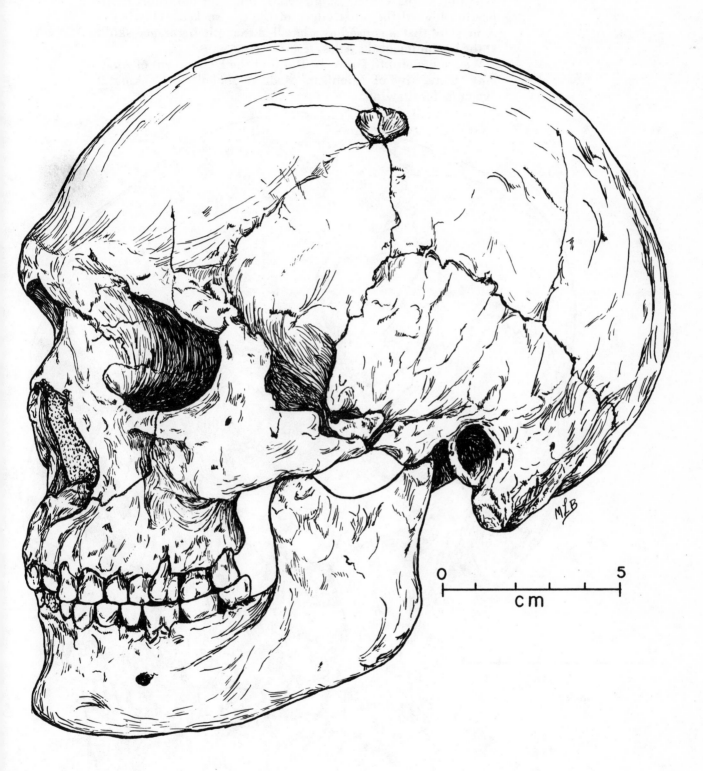

DESIGNATION: *Homo sapiens,* **male**

COMMENT: Pictured is a relatively typical modern male of European extraction. The remnants of the brow ridge occur as a slight swelling in the center (at glabella), thinning out, and disappearing toward the outer edges of the eye sockets. This is the same skull that was used aş a labelled example (reference skull) at the beginning of the Atlas.

We are grateful to Dr. A. R. Burdi of the Department of Anatomy, University of Michigan Medical School, for making it available for drawing.

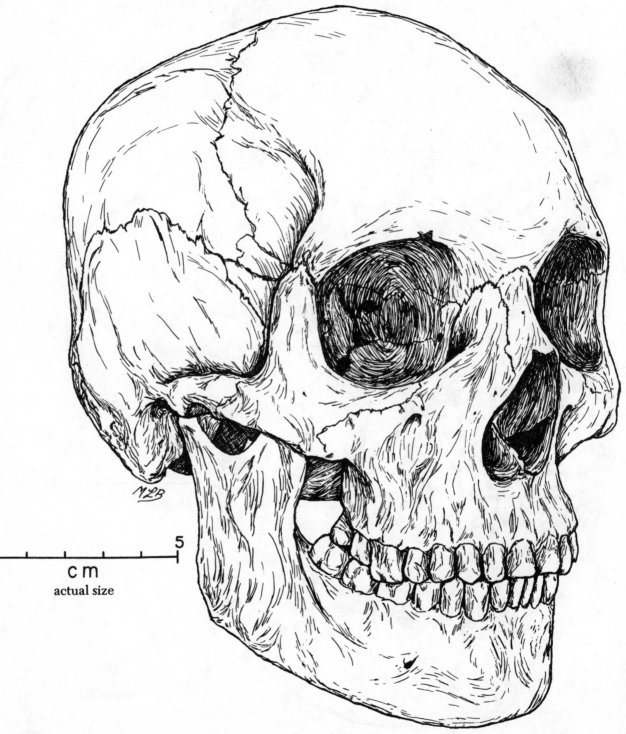

0 _____ 5

c m

actual size

166

DESIGNATION: *Homo sapiens,* **female**

COMMENT: This is a small but relatively typical modern human female. The facial skeleton and dentition of *Homo sapiens* has continued to reduce during the last 10,000 years (following the end of the Pleistocene). This is symbolized in the skull pictured by the absence of the third molar, a phenomenon which has become increasingly common during recent centuries. In many modern individuals the process of dental reduction has gone even farther.

Specimen furnished for drawing courtesy of Dr. A. R. Burdi, Department of Anatomy, University of Michigan Medical School.

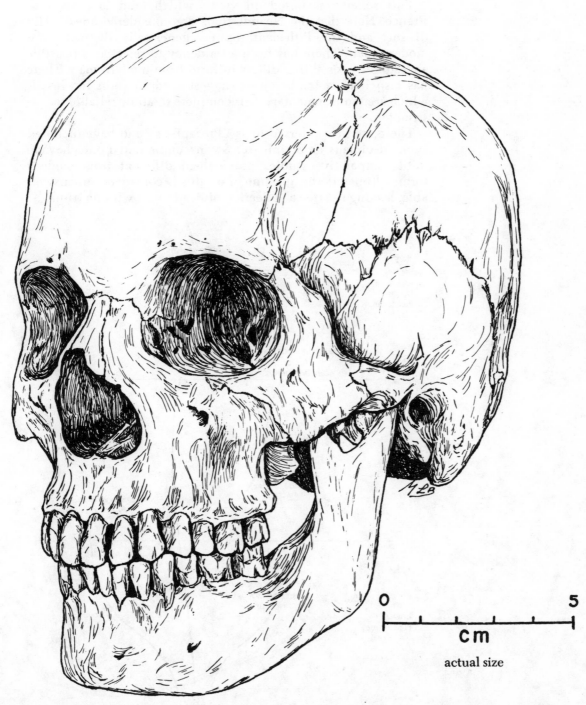

0   5

cm

actual size

Phylogenetic Trees

Any number of phylogenetic trees have been suggested purporting to represent the course of evolution leading up to modern man. From time to time, proponents of one scheme or another change their views so that no rigid formalization is an accurate representation of any major school of thought. There are distinct differences in approach, however, and the diagrams that follow are attempts to capture the different emphases.

SCHEME A

This scheme is based on views which tend to prevail in France. Note that the Neanderthals are considered specifically distinct and the Pithecanthropines generically distinct from modern man. There has been a tendency to reject as a possible ancestor any fossil that differs in form from modern man. There has also been a tendency to suggest modern form for fossils which are too fragmentary or incomplete to allow reliable reconstruction.

These are then referred to as "Presapiens" and have included such specimens as Steinheim, Swanscombe and Fontéchevade until comparative studies prove them different from modern form. Ultimately the question of origins is considered unanswerable, leaving the tree a collection of dead twigs with no trunk.

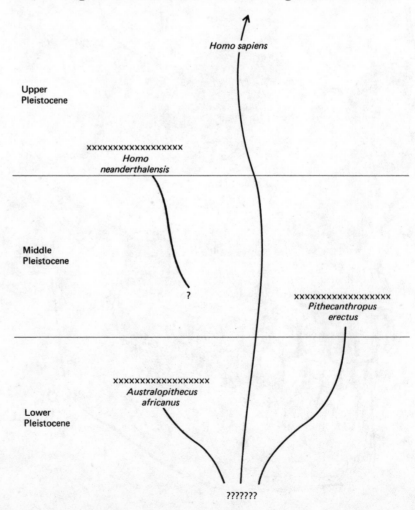

SCHEME B

One version or another of this tree can be identified with Dr. L. S. B. Leakey, the late director of the Centre for Prehistory and Palaeontology at the National Museum of Kenya in Nairobi. Most students regard *"Proconsul"* as *Dryopithecus,* and *"Kenyapithecus"* (at least, the later specimens so identified) as *Ramapithecus,* but, given these changes in designation, agree that this is indeed the course of development that led to true human beings.

Most now also regard *"Homo habilis"* as an invalid designation, the earlier specimens being Australopithecines and the later ones Pithecanthropines. Before his death, Dr. Leakey came to accept the identification of *"Zinjanthropus"* as an Australopithecine and considered the whole group as an extinct side line.

There is some question in this tree about where the Pithecanthropines branched off, but in any case they are considered ancestral to the Neanderthals who are regarded as an extinct side line. As in the French scheme, there is a tendency to give nonmodern forms separate generic and/or specific designations and eliminate them from any role in the ancestry of modern man.

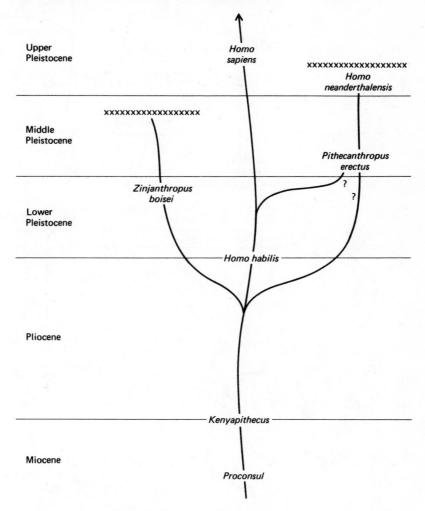

169

SCHEME C

This is the scheme favored by many students of human evolution today. The Australopithecines are split into two groups sometimes considered generically and sometimes only specifically separate. The *"Paranthropus"* group (sometimes regarded simply as *Australopithecus robustus*) includes Australopithecines from Swartkrans, Kromdraai, the "Zinj" specimen from Olduvai Gorge, and perhaps *"Meganthropus"* from Java. They are considered a retarded and poorly adapted form which became extinct by the Middle Pleistocene as a result of competition with more advanced hominids. At a later time level, the same arguments are used in regard to the Neanderthals.

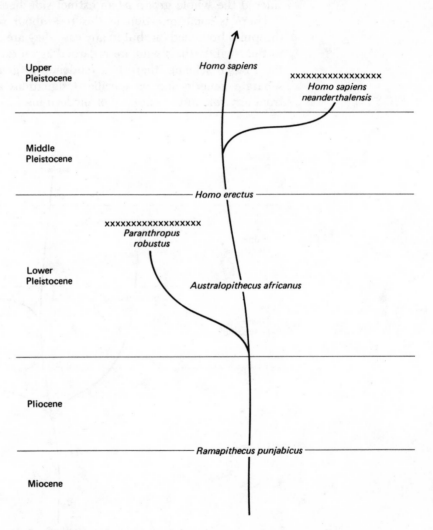

Upper Pleistocene

Homo sapiens

xxxxxxxxxxxxxxxx
Homo sapiens neanderthalensis

Middle Pleistocene

Homo erectus

xxxxxxxxxxxxxxxxx
Paranthropus robustus

Lower Pleistocene

Australopithecus africanus

Pliocene

Ramapithecus punjabicus

Miocene

SCHEME D

Relatively few students accept this picture which was once favored by the senior author of this atlas. The course of human evolution is arranged in four broad stages (Australopithecine, Pithecanthropine, Neanderthal and Modern) each of which is regarded as having evolved directly from the preceding. It has been criticized as being simplistic. Most scholars believe that the course of human evolution has been too complex and that the evidence is too fragmentary to be represented by such a simple scheme.

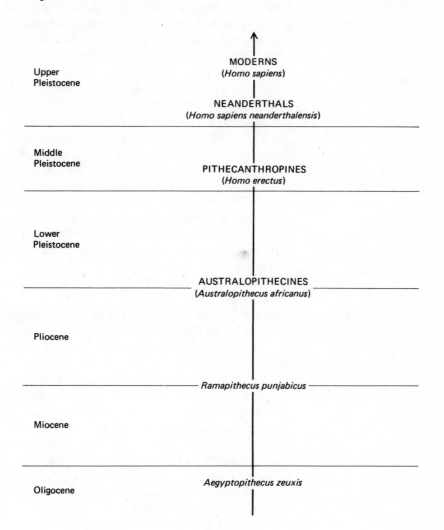

The Hominid Family Tree

Or putting some teeth into generalizations about the course of hominid evolution.

The vertical dimension represents time, and obviously the scale over the past million years is extended when compared to the scale between 1 and 3 million years. The horizontal dimension represents total average tooth size in hundreds of square millimeters. This is calculated by adding up the cross-sectional areas of the upper and lower teeth of individuals to get an average figure for a given population. The dots on the chart, then, represent actual groups for which data are available. The basal dot was constructed from the data published for the Hadar and Laetoli specimens from the Pliocene of East Africa. The dot immediately above that represents all the information available from Sterkfontein and Makapan at the end of the Pliocene in South Africa. Above and to the right of that, the dot represents tooth size at Swartkrans and Kromdraai. The dot labeled *A. boisei* was constructed from the measurements for "Zinj," Peninj, and the robust Australopithecine teeth in the Koobi Fora area east of Lake Turkana. Most anthropologists now agree that this represents a big-toothed branch of the hominid tree that concentrated on a hard-object/plant-food diet and that became extinct before the beginning of the Middle Pleistocene. The *erectus* dot is constructed principally from the Middle Pleistocene Chinese Pithecanthropine teeth from Choukoutien. The lowest dot in the *sapiens* region represents the early Neanderthals from Krapina in Yugoslavia. The dot immediately above and to the left is the late or "classic" Neanderthals of western Europe, and the modern spectrum ranges from big-toothed Australian aborigines on the top right to Europeans and Chinese on the top left. All the rest of modern *Homo sapiens* fall between these extremes, and, if their dots had been included, they would run one into each other without break.

The plot of tooth size and time represents a picture of hominid evolution from a strictly dental point of view. The teeth, however, have responded to the major hominid adaptive shifts and, since they exist in measurably greater abundance than any other part of the skeleton, they can at least give us a framework with which to begin the organization of our family tree. When we use the information available from the rest of the skeletal evidence for biological adaptation and the archeological evidence for cultural adaptation, we can draw the connecting lines that finish the version of the hominid phylogenetic tree that is presented here.

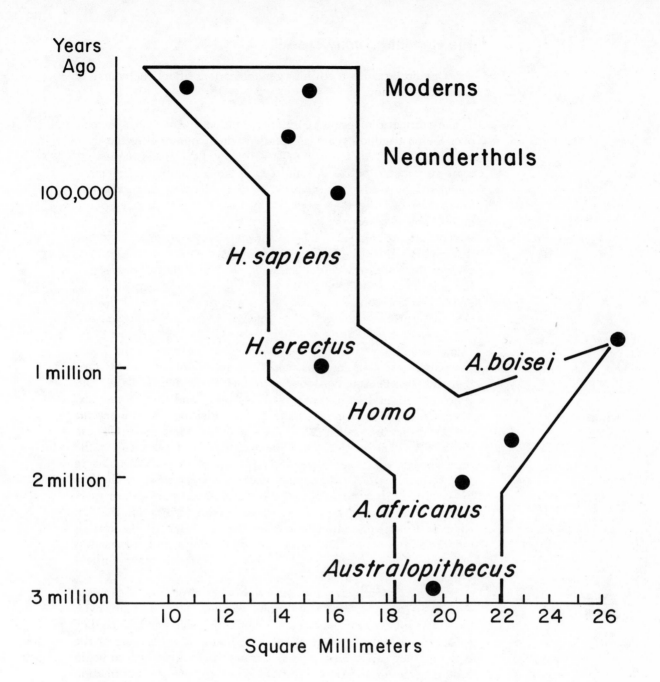

GENERAL REFERENCE SOURCES

Bordes, Francois, 1968, *The Old Stone Age*. Trans. from the French by J. E. Anderson. World University Library, New York: McGraw-Hill Book Company. 255 pp.

Brace, C. Loring, 1979, *The Stages of Human Evolution: Human and Cultural Origins*, 2d ed. Englewood Cliffs, N.J.: Prentice-Hall, Inc. 116 pp.

————, and Ashley Montagu, 1977, *Human Evolution: An Introduction to Biological Anthropology*, 2d ed. New York: Macmillan. 493 pp.

Clark, Grahame, 1977, *World Prehistory in New Perspective*, 3d ed. New York: Cambridge University Press. 574 pp.

Coon, Carleton S., 1962, *The Origin of Races*. New York: Alfred A. Knopf. 724 pp.

Day, Michael H., 1977, *Guide to Fossil Man: A Handbook of Human Palaeontology*, 3d ed. Chicago: University of Chicago Press. 346 pp.

Howell, F. Clark, 1965, *Early Man*. Life Nature Library, New York: Time, Inc. 200 pp.

Hrdlička, Aleš, 1930, *The Skeletal Remains of Early Man*. Smithsonian Miscellaneous Collection, No. 83. Washington, D.C. 379 pp.

Leakey, Richard E., and Roger Lewin, 1977, *Origins: What New Discoveries Reveal about the Emergence of Our Species and Its Possible Future*. New York: Dutton. 264 pp.

Oakley, Kenneth P., and Bernard G. Campbell, 1967, *Catalogue of Fossil Hominids, Part I: Africa*. London: British Museum (Natural History). 128 pp.

Oakley, Kenneth P., Bernard G. Campbell, and Theya Ivitsky Molleson (eds.), 1971, *Catalogue of Fossil Hominids, Part II: Europe*. London: British Museum (Natural History). 379 pp.

———— (eds.), 1975, *Catalogue of Fossil Hominids, Part III: Americas, Asia, Australasia*. London: British Museum (Natural History). 226 pp.

Poirier, Frank E., 1977, *Fossil Evidence: The Human Evolutionary Journey*, 2d ed. St. Louis: Mosby. 342 pp.

Sauer, Norman J., and Terrell W. Phenice, 1977, *Hominid Fossils: An Illustrated Key*, 2d ed. Dubuque, Iowa: W. C. Brown. 169 pp.

Vallois, H-V., and H. L. Movius, 1953, *Catalogue des Hommes Fossiles*. Macon, France: Protat Frères. 318 pp.

INDEX